Conodont biostratigraphy and taxonomy of the Ordovician shelf margin deposits in the Scandinavian Caledonides

JAN AUDUN RASMUSSEN

Rasmussen, J.A. 2001 05 01: Conodont biostratigraphy and taxonomy of the Ordovician shelf margin deposits in the Scandinavian Caledonides. *Fossils and Strata*, No. 48, pp. 1–180. Denmark. ISSN 0300-9491. ISBN: 14-05-16988-5

The Lower and Middle Ordovician (upper Tremadoc – lower Caradoc) succession in the Lower Allochthon of the Scandinavian Caledonides has been sampled with particular emphasis on the limestone intervals, most notably the Lower Ordovician (Arenig – Llanvirn) Stein Formation and the Middle Ordovician Elvdal Formation (Upper Llanvirn – Lower Caradoc). Conodonts were examined in 133 samples from 15 localities situated along the Caledonian front from central south Norway (Valdres) to northwestern central Sweden (Jämtland). A total of 19,286 specimens were recovered, representing 398 morphotypes, which are referred to 45 genera and 102 species. Three genera and four species are new: the genera *Costiconus*, *Minimodus*, *Nordiora* and the species *Microzarkodina corpulenta*, *Minimodus poulseni*, *Nordiora torpensis* and *Triangulodus amabilis*. Because of its ocean-margin habitat, the conodont fauna of the Stein Formation contains several taxa that previously have not been reported from the Baltoscandian platform but are common in the marginal areas of North America. This makes the Stein Formation an effective link for biostratigraphical correlations across the Iapetus Ocean. The conodont fauna differs from the faunas that typify the more proximal, undeformed Baltoscandian platform, with respect to stratigraphic ranges and relative abundance. It is especially distinctive for the upper Arenig – lower Llanvirn time interval. As a consequence, seven new conodont zones have been established for the critical interval.

Key words: Conodonts; biostratigraphy; zonation; systematics; Baltoscandia; platform margin; Caledonides; Lower Allochthon; Tremadoc; Lower Ordovician; Middle Ordovician; Arenig; Llanvirn; Llandeilo; Caradoc; Norway; Sweden.

J.A. Rasmussen, [jar@geus.dk], Department of Stratigraphy, Geological Survey of Denmark and Greenland, Thoravej 8, DK-2400 Copenhagen NV, Denmark; 2nd May 1996; revised 15th June, 1998.

Contents

Introduction

The knowledge of the Ordovician succession in the Caledonian orogenic belt is relatively poor compared to the contemporary autochthonous strata of the Baltoscandian platform as a result of folding, faulting and metamorphism. However, during the last two decades an increasing number of papers have focused on Caledonide palaeontology, sedimentology and structural geology, e.g. Bockelie & Nystuen (1985), Hossack et al. (1985), Nickelsen et al. (1985), Bruton & Harper (1985, 1988); Spjeldnæs (1985); Bruton et al. (1989); Palm et al. (1991) and Stephens et al. (1993), which have led to a better understanding of the areas near the Baltoscandian platform margin facing the Iapetus Ocean.

The most conspicuous Ordovician limestone units in the Lower Allochthon were known as the "Orthoceras Limestone" in the older literature, but were subsequently named the Stein Formation (Lower Ordovician) and the Elvdal Formation (Middle Ordovician) by Rasmussen & Bruton (1994). The limestone units have undergone extensive pressure-dissolution due to compaction, and most of the macrofauna has been damaged. In contrast, conodonts recovered from these low-grade metamorphic limestones were fractured, but in general, most were adequately preserved for species identification, and thus available for age determination.

Previously, conodonts from the Ordovician sediments of the Lower Allochthon have been reported and discussed by Bergström (1971, 1973, 1986, 1988: include Jämtland [Sweden]; Kohut (1972: Steinsodden, Herram [Norway]); Bergström et al. (1974: Andersön [Sweden]); Rasmussen & Stouge (1989: Høyberget [Norway]) and Repetski in Bruton et al. (1989: Grøslii [Norway]). Bergström (1971, 1979, 1997) described Ordovician conodonts from the Upper Allochthon of the Norwegian Caledonides. The present paper, however, is the first attempt to describe the faunal succession through the Stein and Elvdal formations with regard to systematic palaeontology and biostratigraphy. The conodont palaeoecology of the studied sections and its relation to the relative sea-level changes were described by Rasmussen & Stouge (1995), and aspects of the conodont palaeobiogeography of the areas described in this paper and selected areas surrounding the Iapetus Ocean by Rasmussen (1997).

Geological setting

Deposition of the Baltoscandian Ordovician took place in a large, epicontinental sea, which extended from Russia (Moscow Basin), Ukraine and the Baltic countries in the east to Jämtland in the west, and from Lapland in the north to Scania in the south. The relatively nearshore areas of the stable platform (St. Petersburg – Estonia – Ukraine – eastern Sweden) were characterised by extensive carbonate deposition, whereas a higher content of siliciclastic mud typified the more distal platform areas towards the west (Scania, Bornholm, western Västergötland, Oslo). The Baltoscandian platform has been subdivided into distinct, chiefly SE – NW oriented, belts of different lithologies (Männil 1966) and faunas, referred to as confacies (Jaanusson 1976). The depositional setting changes from shallow marine, glauconitic limestones and terrigenous sediments in the east (North Estonian Confacies, Lithuanian Confacies, Moscow Basin) to deeper, but still relatively shallow, marine limestones within the Central Baltoscandian Confacies Belt (e.g. Öland, Östergötland, Siljan region).

The Scanian Confacies Belt and the Oslo belts, consist of mainly graptolitic shales with minor limestone formations. The Ordovician of the Oslo region is dominated by dark shales with numerous, thin, nodular

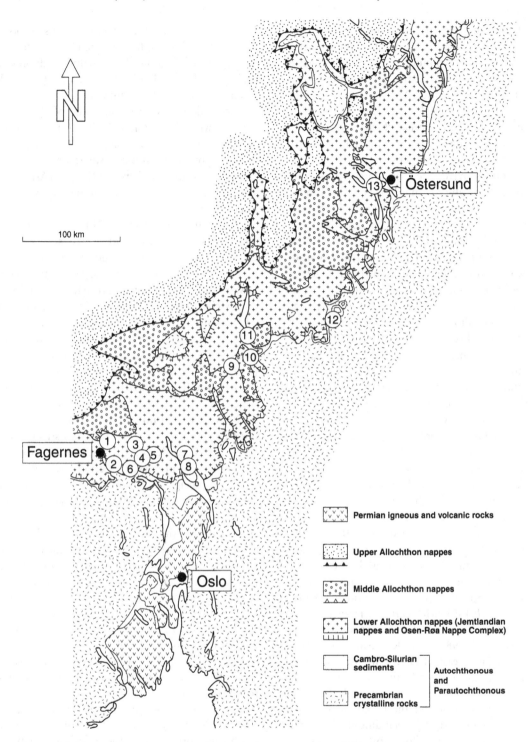

Fig. 1. Location map modified from Rasmussen & Bruton (1994). □1 = Grøslii, □2 = Hestekinn, □3 = Jøronlia, □4 = Røste, □5 = Haugnes, □6 = Skogstad, □7 = Steinsodden and Steinsholmen, □8 = Herram, □9 = Høyberget, □10 = Røskdalsknappen (Engerdal), □11 = Sorken, □12 = Glöte, □13 = Andersön.

limestone horizons (Størmer 1953). Bockelie (1978) demonstrated that the Oslo region may be separated into four distinct semi-concentric faunal belts, which indicate gradually shallower conditions towards the west and northwest from Oslo. The Oslo region sediments contain on average 50–60% terrigenous components, which is considerably more than the

5–10% typical of the sediments of the inner shelf (Estonia) (Bockelie 1978).

The Scanian Confacies Belt and the southern part of the Oslo belts is autochthonous, whereas its northern part (north of Slemmestad) is composed of parautochthonous sediments, which were deformed during the Scandian phase (Silurian) of the Caledonian

Orogeny. The Cambrian to lowermost Ordovician Alum Shale Formation served as décollement surface.

The Scandinavian Caledonides are exposed over a length of 1800 km from Stavanger (southwestern Norway) through Hamar and Östersund to Vadsø in northern Norway. The Caledonian deformation produced a variety of thrust sheets, which were derived from the west and northwest and moved onto the undeformed cover of the Baltic craton.

The Caledonides have been separated into four tectonostratigraphic units: The Lower Allochthon, Middle Allochthon, Upper Allochthon and Uppermost Allochthon (Roberts & Gee 1985). The Lower Allochthon, Middle Allochthon and parts of the Upper Allochthon represent the tectonically shortened margin of Baltica, while most of the Upper Allochthon together with the Uppermost Allochthon have been referred to exotic terranes outboard of the Baltic craton (Stephens 1988). The Lower Allochthon nappes are completely dominated by latest Precambrian and Early Palaeozoic sedimentary successions, and include, for example, the predominantly Vendian siliciclastic mud- and sandstones and conglomerates, previously referred to as "the sparagmites". Cambro-Ordovician shales and Ordovician limestones and greywackes (to the west) are also common. The sediments were telescoped by folds and listric reverse-faults during the Scandian phase of the orogeny, which means that the most common structural features within the Lower Allochthon are listric faults (high-angle), thrusts (low-angle) and imbrications (Bockelie and Nystuen 1985).

The limestone units investigated here, that is the upper Tremadoc Bjørkåsholmen Formation, the Arenig–Llanvirn Stein Formation and the upper Llanvirn–lower Caradoc Elvdal Formation, occur in the Osen-Røa Nappe Complex in Norway (Nystuen 1981; Hossack *et al.* 1985, with earlier references), and in the Jemtlandian Nappes in Sweden (Asklund 1960; Gee 1975; Gee *et al.* 1985). The Røa Nappe is thrust upon the Osen Nappe east of Lake Femunden (central east Norway), but the nappes form one tectonic unit west of the lake (Bockelie & Nystuen 1985). In the area between East Jotunheimen and the Caledonian nappe front at Gjøvik–Torpa–Aurdal (including the localities 1–6 at Fig. 1), the Lower Allochthon comprises, in ascending order, the Osen and the Synnfjell nappes (Hossack *et al.* 1985). The Haugnes, Røste, Jøronlia, Skogstad and Hestekinn sections are part of the Aurdal duplex, which are bordered by the Aurdal Thrust (floor thrust) and the Synnfjell Thrust (roof thrust). The Aurdal duplex comprises the upper Vendian–(?) lower Cambrian siliciclastic Vangsås Formation and the overlying Cambro-Ordovician sediments of the frontal part of the Osen Nappe. The tectonically overlying Synnfjell duplex is bordered by the Synnfjell Thrust

(floor thrust) and the Mellane Thrust (roof thrust) (Hossack *et al.* 1985, fig. 5). The Grøslii section belongs to this tectonic unit. The Synnfjell duplex consists mainly of Upper Precambrian to Middle Ordovician (Llanvirn) siliciclastic sediments.

The Andersön sections are part of the Jemtlandian Nappes, whereas the Glöte section at Härjedalen (Fig. 1) belongs to the autochthonous sequence, which is overlain by Precambrian siliciclastic rocks of the Lower Allochthon Vemdal Nappe (Asklund 1933). The Vemdal Nappe is a lateral equivalent to the Osen Nappe.

The former "*Orthoceras* Limestone" (e.g. Størmer 1953) was shown to include two separate limestone units by Rasmussen & Stouge (1989). In south central Norway, the Stein Formation separates the shales of the Lower Ordovician Tøyen Formation and the Middle Ordovician Elnes Formation in the Aurdal duplex, whereas the Ordovician succession within the structurally overlying Synnfjell duplex is dominated by siliciclastic sediments of the Ørnberget Formation (Figs. 2, 3) (Nickelsen *et al.* 1985; Hossack *et al.* 1985).

The Middle Ordovician (Upper Llanvirn (Llandeilo)–?Caradoc) Elvdal Formation was observed only in east central Norway close to the Swedish border (Fig. 1: 9–11). The unit was strongly deformed and fragmented during the Caledonian orogeny, but a relatively well-preserved section was discovered at Høyberget (Fig. 1: 9) during the field-season 1993 (Rasmussen & Bruton 1994). The "Biseriata Limestone" at Andersön, Jämtland is regarded as the lateral equivalent to the Elvdal Formation. The "Biseriata Limestone", however, is less argillaceous and was deposited in a more shallow water environment than the Elvdal Formation, resulting in that the two units have a different erosional surface pattern (compare Figs. 16 and 22A, B).

The Early Ordovician (Tremadoc) Bjørkåsholmen Formation is characterised by a very uniform development across the Baltoscandian platform, which indicates that the platform during this interval was relatively plane, with a minimal relief. However, the unit disappears towards the platform margin, and is replaced by small, nodular limestone beds within the Ørnberget Formation of the Synnfjell duplex.

Stratigraphy and lithology

The former "*Orthoceras* Limestone" of the Lower Allochton were described in detail by Rasmussen & Bruton (1994), and were named the Stein and Elvdal formations, respectively. These two units together with a few sections of the Tremadoc Bjørkåsholmen Formation (previously "*Ceratopyge* Limestone") were

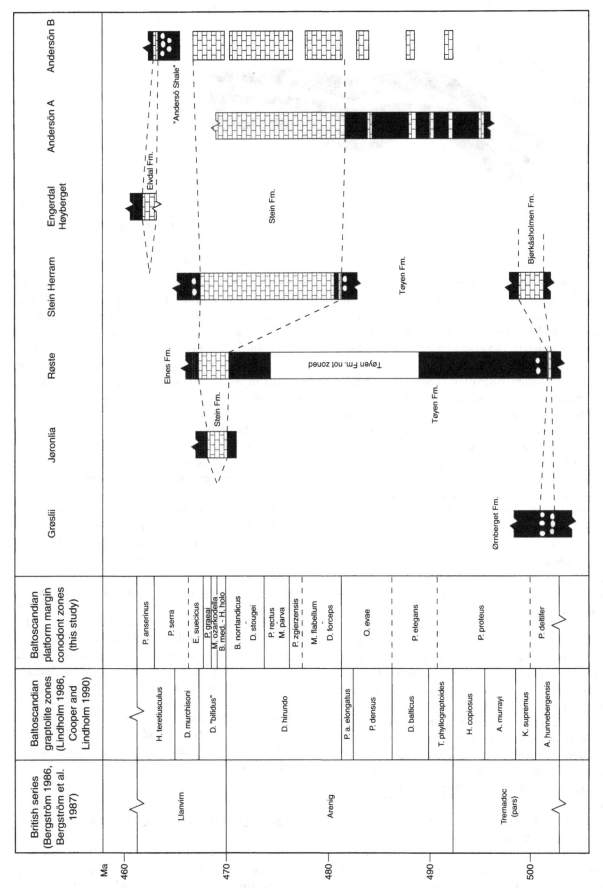

Fig. 2. Stratigraphic position of the investigated sections. Note the diachronous nature of the Stein Formation.

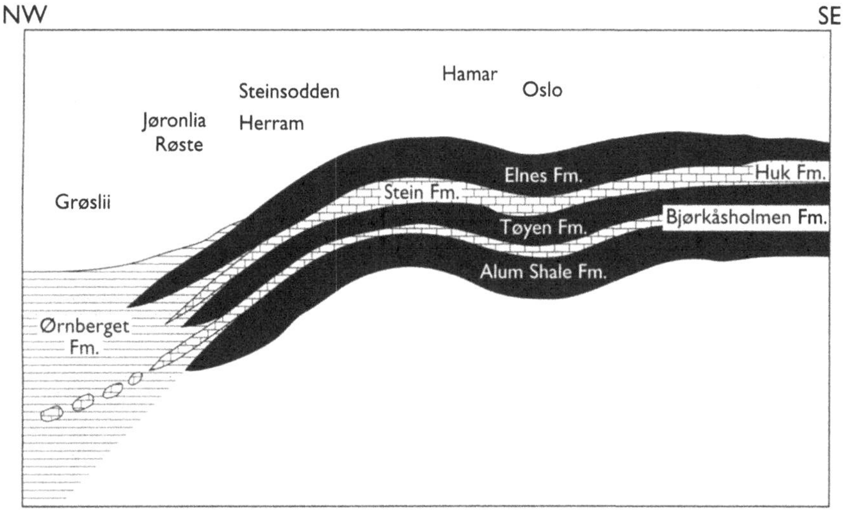

Fig. 3. Palinspastic reconstruction of the Lower Ordovician Baltoscandian margin areas, southern central Norway (not to scale). The relative distance from Hamar to Herram and Steinsodden is strongly understated.

sampled and investigated in the present study. The Ordovician lithostratigraphy of the Oslo Region, including the Lower Allochthon Ringsaker District, was revised and described by Owen *et al.* (1990). Several of the units do also exist in the Lower Allochton of the Caledonian foldbelt.

The Bjørkåsholmen Formation (Owen *et al.* 1990), which separates the shales of the Cambrian to lowermost Ordovician Alum Shale Formation from the uppermost Tremadoc and Arenig Tøyen Formation, becomes thinner towards the west (Fig. 3). It consists of alternating limestone beds and shales at Steinsodden, where it is 1.5 m thick. At Røste the thickness has decreased to 0.9 m, while the corresponding unit at Grøslii consists of shales of the Ørnberget Formation with one or more intercalating thin limestone nodule horizons.

The succeeding siliciclastic shales of the Tøyen Formation commonly have a slaty cleavage. The thickness approximates to 14 m at Røste, while it is unknown at the other localities investigated in this study.

The upper Arenig–lower Llanvirn Stein Formation (Skjeseth 1963) is dominated by grey, massive mud- and wackestone *sensu* Dunham (1962). Pressure dissolution seams containing residual clay form a characteristic reticulate pattern. Skeletal debris of trilobites, brachiopods and more rarely gastropods is visible in thin sections from the Stein Formation at Steinsodden. The formation is characterised by beds of argillaceous limestone alternating with beds of more pure limestone. Distinct bedding planes are rare. Thickness of the individual beds is usually 10–30 cm. Clastic silt beds become progressively more common towards the west.

The Stein Formation comprises the Herram and Steinsholmen members. At Herram, the c. 1 m thick Herram Member consists of grey shales with 8 nodule horizons. The nodules consist of biogenic, marly limestone. The base of the succeeding Steinsholmen Member is placed at the base of the lowest coherent limestone bed at Herram. The lower c. 1.5 m of the Steinsholmen Member consists of shales and intercalating limestone beds. The succeeding massive part consists of four principal lithofacies at Ringsaker and Andersön, whereas it is restricted to only one lithofacies west of Lake Mjøsa.

Siliciclastic shales of the Llanvirn Elnes Formation succeed the Stein Formation. It consists of alternating muddy and silty/sandy beds near Røste, which may indicate a slope location. The Elnes Formation contains a few nodular horizons close to the base of the formation at Steinsodden. The basal part of equivalent "Andersö Shale" at Andersön is apparently younger (*H. teretiusculus* Zone) than the lower part of the Elnes Formation (*D. murchisoni* Zone).

The upper Llanvirn–lower Caradoc Elvdal Formation was observed in the Engerdal area of eastern Norway. The "Biseriata Limestone" from Andersön, Jämtland is the northeastern equivalent unit, and is probably a member of the Elvdal Formation. The Elvdal Formation consists of medium grey, weakly-bedded, biogenic, argillaceous lime mudstone with a characteristic irregular, reticulate weathered surface, similar to that of the Stein Formation.

The Elvdal Formation is succeeded by an unnamed shale unit at Høyberget and Engerdal. The unit is poorly dated (Rasmussen & Stouge 1989), but is probably Caradoc in age.

Description of localities

Thirteen localities situated along the Caledonian front from Fagernes and Torpa in south-central Norway, through Ringsaker and the Engerdal–Drevsjø area in central East Norway, to Jämtland in northern, central Sweden were sampled (Fig. 1). The localities are described from southwest towards northeast. The most comprehensive and stratigraphically most important localities are the Steinsodden, Andersön-A and Andersön-B sections.

The Tremadoc Bjørkåsholmen Formation was investigated at Steinsodden (Fig. 1: 7) and Røste (Fig. 1: 4). The Arenig–Llanvirn Stein Formation was sampled at Hestekinn (Fig. 1: 2), Jøronlia (Fig. 1: 3), Røste (Fig. 1: 4), Haugnes (Fig. 1: 5), Skogstad (Fig. 1: 6), Steinsodden (Fig. 1: 7), Herram (Fig. 1: 8) and Andersön (Fig. 1: 13). The Elvdal Formation (Llanvirn–Caradoc) seems to be restricted to the easternmost Norway, at Høyberget (Fig. 1: 9), Røskdalsknappen (Fig. 1: 10), Sorken (Fig. 1: 11). It is possible, however, that the lateral equivalent limestone unit at Andersön (Fig. 1: 13), the "Biseriata Limestone", also should be regarded as a member of the Elvdal Formation.

Neither Strand (1954) nor the present author did observe outcrops comprising the Stein Formation west of Hestekinn, Bruflat, but contemporary thin limestone beds may occur at Grøslii (Bruton *et al.* 1989).

Grøslii, Norway

Location. – The Grøslii section (Fig. 1: 1) is situated at Grøslii seter on the eastern slope of Skardåsen (1071 m), Valdres, some 15 km north of Leira (map sheet Fullsenn 1717 III, UTM NN173703).

Characteristics. – The locality is an inverted (younging downwards) slope section north of Øygården, Grøslii. It consists entirely of Cambro-Ordovician slates with minor intercalations of limestone (nodules and thin beds), and siliciclastic silt- and sandstones. The sediments are generally well exposed. The conodont bearing part is orientated 78°/30°NNW, whereas the overlying, but stratigraphically older interval just above the deformed zone is orientated 170°/26°W.

Previous work. – The Valdres area "Nordre Etnedal" was mapped and described by Strand (1938). The structural geology and stratigraphy of the area were revised and outlined by Nickelsen *et al.* (1985), who concluded that the Solheim Slate Member extends from the Tremadoc *Rhabdinopora flabelliforme* graptolite zone to the lower Llanvirn with the youngest limestone facies found in the upper Tremadoc. Bjørlykke (1905) and Williams (1984) described the middle Arenig–lower Llanvirn graptolite fauna from equivalent beds in Gausdal.

Bruton *et al.* (1989) discussed the Caledonian trilobite, brachiopod and conodont faunas and their stratigraphical significance. They illustrated the Grøslii

faunas and Repetski (in Bruton *et al.* 1989) listed and discussed a relatively diverse conodont fauna from the Solheim Member, which correlates with the material presented here.

Stratigraphic succession. – The studied interval occurs from 940 m to about 950 m (altitude) and includes a slate sequence about 30 m thick of the Solheim Member of the Ørnberget Formation (Nickelsen *et al.* 1985) dipping towards the north. The sequence is only insignificantly internally deformed. The overturned sequence is succeeded by a strongly deformed thrust zone about 15 m thick which again is overlain by slates with intercalating silt- and sandstones.

The study concentrated on small nodular horizons in the lowest, but stratigraphically youngest, Solheim Slate sequence. One of the nodular horizons correlates biostratigraphically with the Bjørkåsholmen Formation of the Oslo Region. It contained 80% $CaCO_3$ and almost 1400 ppm strontium (Appendix 1). This is about three times the amount of strontium that characterises the Stein Formation and more than four times the amount within the Elvdal Formation.

Conodont biostratigraphy. – The three analysed samples contained a sparse conodont fauna dominated by *Paltodus* cf. *deltifer* and *Paroistodus numarcuatus* and correlates with the *P. deltifer* Zone of Lindström (1971).

Hestekinn, Norway

Location. – The locality is situated about 3 km west of the village Bruflat, map sheet Bruflat 1716 I, UTM NN323498 (Fig. 1: 2).

Characteristics. – The locality is composed of several small limestone exposures arranged in narrow east-west oriented ridges in the forest northwest of Hestekinn Farm. The section is the westernmost known *"in situ"* Stein Formation exposure (Figs. 4, 5). The beds dip towards the north (88°/37°N).

Previous work. – Strand (1954) mapped and described the Aurdal area. He visited Hestekinn and noted that the limestone is similar to "Orthocerkalken" (= the Stein Formation) north of the Oslo Region. The thickness was estimated to up to 12 m (Strand 1954, p. 33), which is a little more than that calculated by the present author. Bjørlykke & Skålvoll (1979) published the Bruflat geological map sheet (Bruflat 1716, 1) which includes the Hestekinn locality.

Stratigraphic succession. – The Stein Formation is composed of impure, grey, reticulate mudstone, but is often covered by soil. The thickest continuous section

Fig. 4. The Stein Formation at Hestekinn.

is approximately 4.3 m. The thickness of the Stein Formation is unknown, but it is at least 4.3 m. The average $CaCO_3$ content is 56%. Grey shales of the Tøyen Formation underlie the Stein Formation.

Conodont biostratigraphy. – The two conodont bearing samples correlate with the *B. medius – H. holodentata* Zone. Sample 97706 is from a level close to the base of the Stein Formation.

Jøronlia, Norway

Location. – The section is situated 500 m SE of the small village Jøronlia in Nord-Torpa some 25 km north of Dokka (map sheet Bruflat 1716 I, UTM NN505617) (Fig. 1: 3).

Characteristics. – Most of the Stein Formation exposures in Nord-Torpa are placed along the rivulet Gjerda. The sections are easily accessible but the stratigraphic succession is disturbed due to folding and faulting. The most complete section was observed in the forest between Gjerda and Jøronlia (Figs. 6, 7, 8). The shales dip 72°/26°NW.

Previous work. – The area was mapped by Münster (1901) and Bjørlykke & Skålvoll (1979).

Stratigraphic succession. – The Stein Formation is rather impure but is also characterised here by the typical reticulate pattern. It consists of dark grey limestone beds alternating with siliciclastic silt beds. The $CaCO_3$ content varies from 40 – 70% in the limestone layers to 4 – 17% in the silty beds. The thickness is 5.8 m. Dark grey slate, which is probably part of the Tøyen Formation, underlies the Stein Formation. The Stein Formation is succeeded by grey, partly silty, slates of the Elnes Formation.

Conodont biostratigraphy. – The Stein Formation correlates with the *B. medius – H. holodentata* Zone and the *P. graeai* Zone.

Røste, Norway

Location. – The locality is situated at Røste Farm, Aust-Torpa, c. 14 km north of Dokka (map sheet Dokka 1816 IV, UTM NN614583) (Fig. 1: 4).

Characteristics. – The locality is a beautifully exposed roadcut, where the road cuts a small, east-west oriented ridge made up by the Stein Formation

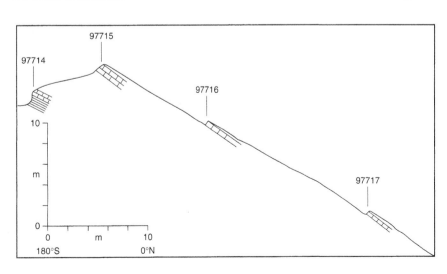

Fig. 5. Hestekinn, indicating the position of the studied conodont samples from the Stein Formation.

Fig. 6. The middle part of the Stein Formation at Jøronlia. The length of the ruler is 1 m.

Legend

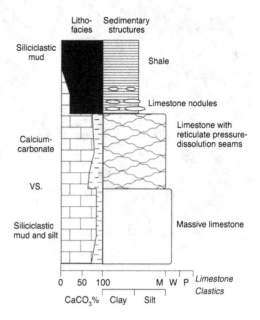

Fig. 8. Legend showing the most significant lithological and sedimentary characteristics used in figures. The calcium-carbonate content was determined with an atomic absorption spectrophotometer (Appendix 1). M, W and P of the sedimentary log refer to mudstone, wackestone and packstone, respectively (Dunham 1962). The numbers to the left indicate the thickness in metres.

(Figs. 9, 10). The dip varies between 112°/48°SSW and 122°/52°SSW.

Previous work. – The area was mapped and described by Münster (1901) and A. Bjørlykke (1973, 1979). The former mentioned the presence of Tremadoc dark shales ("3a") near Røste.

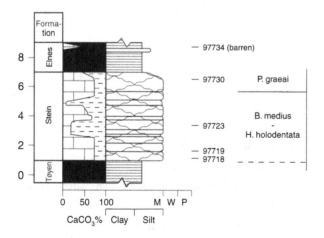

Fig. 7. Stratigraphic column for the Jøronlia section. The conodont zones are indicated to the right. See Fig. 8 for legend.

Stratigraphic succession. – The stratigraphically oldest exposed unit is the Bjørkåsholmen Formation. It is 0.9 m thick and consists of three nodular horizons overlain by two 20 cm thick limestone beds. The unit contains *Ceratopyge forficula* Sars (trilobites determined by Jan Ove Ebbestad, University of Uppsala). A more than 15 m thick shale sequence of the Tøyen Formation succeeds the Bjørkåsholmen Formation. Three nodular horizons occur in the interval 6–6.5 m above the base and one nodular horizon 13 m above the base of the formation. The latter is composed of relatively large limestone lenses, up to 1.7 m long and 0.4 m thick.

The Stein Formation is 7.0 m thick and consists of biogenic, grey, reticulate mud- and wackestone. Cephalopods occur throughout the formation, but are especially common within the upper 0.8 m. Siliciclastic silt beds occur sporadically in the middle part of the formation. The $CaCO_3$ content generally varies between 45 and 72%, but is somewhat higher in the upper 0.8 m (82%) (Appendix 1).

Dark grey shales of the Elnes Formation succeed the Stein Formation. The Elnes Formation is not exposed but may be easily uncovered by digging.

Conodont biostratigraphy. – The lower nodular limestone level within the Tøyen Formation (6–6.5 m

(A)

(B)

Fig. 9. The Røste section. □Fig. 9A: The southwest dipping (towards the right) Stein Formation at Røste. □Fig. 9B: The Bjørkåsholmen section (lower left corner) is suceeded by the Tøyen Formation and the Stein Formation.

above the base) contained conodonts indicative of the *P. deltifer* Zone. The upper nodular horizon was barren. The presence of *P. deltifer* Zone conodonts above the Bjørkåsholmen Formation may indicate that this part of the section is tectonically disturbed.

The Stein Formation correlates with the *B. medius – H. holodentata*, *M. ozarkodella*, *P. graeai* (?) and *E. suecicus* zones.

Haugnes, Norway

Location. – The section is located near the Haugnes Farm in Aust-Torpa, Snertingdal, map sheet Dokka 1816 IV, UTM NN643582 (Fig. 1: 5).

Characteristics. – The locality is a tectonically deformed road section, in total about 70 m long and 1 – 1.5 m high. The dip varies from 18°/36°WNW to 58°/62°NW.

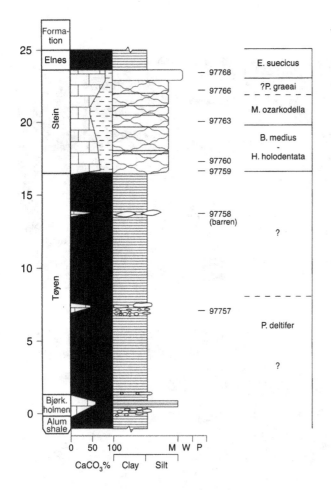

Fig. 10. Stratigraphic column for the Røste section. The conodont zones are indicated to the right. See Fig. 8 for legend.

Previous work. – The geology of the Snertingdal area was mapped and described by Münster (1891, 1900) and A. Bjørlykke (1973, 1979). The strike and dip of the Haugnes section was marked on the geological map DOKKA 1816,4 (Bjørlykke 1973).

The Haugnes section contains a relatively rich early Llanvirn nonarticulated brachiopod fauna, which is distinct from the faunas known from the autochthonous Baltoscandian areas (Harper & Rasmussen 1997).

Stratigraphic succession. – The northwest dipping section is apparently about 24 m thick, and includes the Stein Formation and c. 3 m of the Elnes Formation. However, the section is deformed and partly repeated, and the correct thickness is difficult to estimate. At the locality Røste (see below) situated 3 km farther west, the thickness is 7 m. The section is composed of impure, reticulate limestone.

Conodont biostratigraphy. – Two samples were investigated for conodonts, representing the interval from the uppermost part of the *B. norrlandicus – D. stougei* Zone to the *B. medius – H. holodentata* Zone.

Skogstad, Norway

Location. – The locality is a roadcut near Skogstad Farm, c. 8 km ENE of Dokka (map sheet Dokka 1816 IV, UTM NN647481) (Fig. 1: 6).

Characteristics. – The badly preserved roadcut is about 35 m long and up to 2.5 m high. It is covered through most of the interval from 20 to 35 m. Dip is 106°/75°S.

Previous work. – The area was mapped and described by Münster (1901) and A. Bjørlykke (1973, 1979).

Stratigraphic succession. – The section is composed of limestones belonging to the Stein Formation. The lower 9 m consists of alternating reticulate and more homogenous beds. The individual beds are 5–15 cm thick. The 9–11 m interval is medium reticulate and partly nodular, while the interval from 11 to 20 m is strongly reticulate without well-defined beds. The exposed parts of the 25–35 m interval is composed of more pure, non-reticulate limestone.

A significant thrust cuts the middle part of the section. Secondary cleavage is distinct in the 9–20 m interval.

Conodont biostratigraphy. – Four 1.5 kg samples were analysed for conodonts (0.05 m, 8.3 m, 25.7 m and 34.5 m). The conodont yield was extremely low. The only sample that contained a stratigraphically significant fauna was the 25.7 m sample, which correlates with the *B. medius – H. holodentata* Zone. The lowermost sample contained *Protopanderodus rectus*.

Steinsodden, Norway

Location. – The locality is situated at Steinsodden, Moelv, map sheet Gjøvik 1816 I, UTM NN919539 (Fig. 1: 7). The locality became a nature preservation area in 1985. In general, the fold axis is oriented c. 280° dipping weakly towards the WNW. The beds at the Steinsodden main section are vertical, oriented 90°/90°. The sediments at the "Steinsodden south" section are oriented 115°/58°SW.

Characteristics. – The locality includes the main section on the outer part of Steinsodden (Fig. 11A–C) peninsula, the "Steinsodden south" section on the southern part of Steinsodden and the small island Steinsholmen c. 150 m WSW of the main section.

The main section comprises all the massive part of the Steinsholmen Member of the Stein Formation (c. 42 m in thickness), while the southern section includes the upper part of the Steinsholmen Member and the lower part of the succeeding Elnes Formation (in total about 8 m). The uppermost part of the alternating shale/limestone interval characterising the lower part of the Steinsholmen

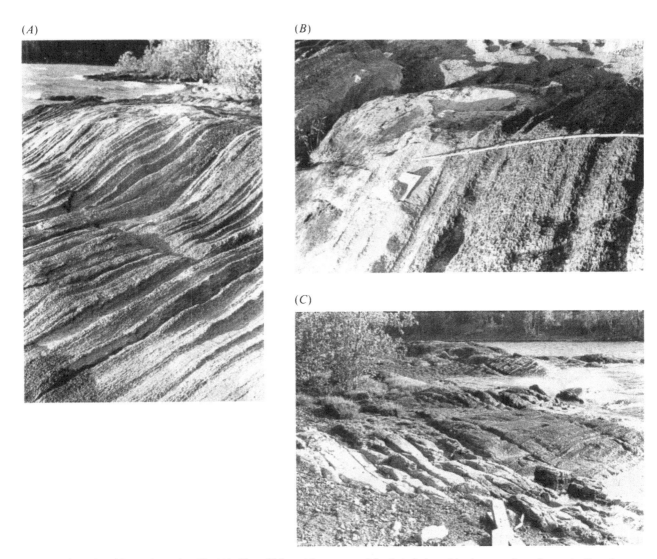

Fig. 11. The Steinsodden main section. Fig. 11A: The middle and lower part of the Steinsholmen Member seen from the south. Note the vertical bedding and the reticulate, erosional pattern. Fig. 11B: Boundary between lithofacies C (left) and lithofacies A in the middle part of the Steinsholmen Member. Fig. 11C: The lower part part of the Steinsholmen Member seen from north-east. The small island of Steinsholmen is visible in the upper part of the picture.

Member is present in the core of the anticline at the southwestern edge of Steinsholmen. The total thickness of the Steinsholmen Member is about 45 m at Steinsodden (Fig. 12). Thus, the total thickness of the Stein Formation in Ringsaker may be estimated at about 47.5 m including the Herram section.

The sediments are easily accessible in the main section as well as the two subsections.

Previous work. – Strand (1929) reported *Ceratopyge forficula*, *Niobe insignis* and *Megalaspis stenorachis?* from the lower part of the Tøyen Formation ("*Phyllograptus* Shale, 3b") at Steinsodden and "*Ogygio dilatata*" from the black shales overlying the Stein Formation (= the Elnes Formation). He estimated the thickness of

the Stein Formation ("Ortoceras Limestone, 3c") to 40 m. It was noted by Størmer (1953) that cephalopods are common in the upper 10 – 15 m of the Stein Formation ("Orthoceras Limestone, 3c").

The Ringsaker area was mapped and described by Skjeseth (1963), who introduced the name "Stein Limestone" for the predominantly massive part of the previous "Orthoceras limestone" in the Ringsaker district.

Conodonts from the Stein Formation were first reported by Kohut (1972). He correlated the unit with the *B. navis*, *P. originalis* and *A. variabilis* zones, but noted that the presence of *Protopanderodus graeai* in the level about 39 m above the base indicates that the uppermost part of the section probably correlates with a level close to

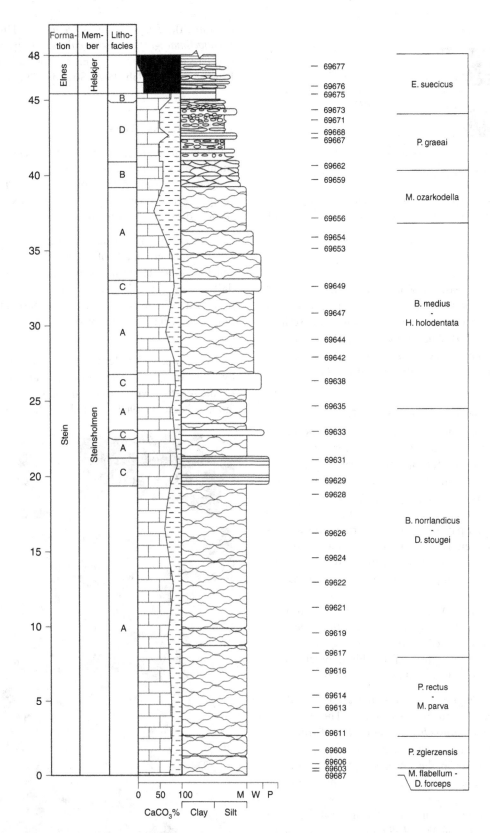

Fig. 12. Stratigraphic column for the Stein Formation at Steinsodden. The conodont zones are indicated to the right. The letters A, B, C, D refer to the observed lithofacies. See Fig. 8 for legend.

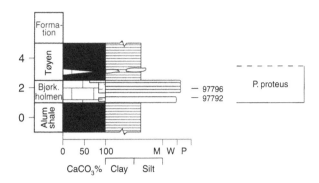

Fig. 13. Stratigraphic column of the Bjørkåsholmen Formation at Steinsodden. Conodont zones are indicated to the right. See Fig. 8 for legend.

the base of the Valastean. Kohut's correlations fit well with the correlations presented here. The Gjøvik map sheet including Steinsodden was mapped and described by Bjørlykke (1979). The lithostratigraphic framework of the "Stein Limestone" was revised by Owen *et al.* (1990) and Rasmussen & Bruton (1994) (see locality Herram).

Stratigraphic succession. – The Alum Shale Formation and the Bjørkåsholmen Formation are well exposed on the northern edge of Steinsodden. The former was not studied in detail but the Bjørkåsholmen Formation is 1.5 m thick and consists of 15 limestone beds and nodule horizons alternating with grey shales (Fig. 13). The main constituent is grey wackestone.

The lowermost part of the Tøyen Formation succeeds the Bjørkåsholmen Formation in the anticline in the north-western edge of Steinsodden. Only loose shale pieces of the overlying part of the Tøyen Formation were observed.

The Steinsholmen Member of the Stein Formation is dominated by grey, massive, biomicritic limestone which may be classified as predominantly mud- and wackestone. Skeletal debris of trilobites, brachiopods and rarely gastropods is visible in thin sections.

Conodont biostratigraphy. – Two samples from the Bjørkåsholmen Formation were analysed. The fauna is dominated by *Paroistodus numarcuatus*, *Paroistodus proteus* and *Paltodus* cf. *subaequalis*, while few specimens of *P. deltifer* occur in the lower sample. The samples correlate with the lower part of the *P. proteus* Zone (Tremadoc).

The Stein Formation includes the interval from the uppermost part of the *M. flabellum* – *D. forceps* Zone to the lower part of the *E. suecicus* Zone (upper Arenig – lower Llanvirn).

Unlike the Stein Formation, the succeeding Elnes Formation is rich in macrofossils and *Ogygiocaris dilatata*

(Brünnich), *Didymograptus geminus* (Hisinger) and *Pseudoclimacograptus scharenbergi* (Lapworth) have been recorded from the Elnes Formation ("4a") at Ringsaker (Holtedahl 1909; Strand 1929; Skjeseth 1963).

Herram, Norway

Location. – The investigated section (Fig. 14) is in a small valley some 200 m west of Herram Farm, Ringsaker, about 7 km southwest of Brummundal (map sheet Hamar 1916 IV, UTM NN989485). The section dips weakly towards the north.

Characteristics. – The locality is a slope-section situated on the northern flank of the small east-west oriented valley behind Herram Farm. The section comprises dark shales of the Tøyen Formation and the shale and limestone horizons of the basal part of the Stein Formation (Fig. 15). The succession is well-exposed and easily accessible.

Previous work. – The Herram section was first described by Skjeseth (1952). He illustrated and described the profile together with parts of the graptolite and trilobite faunas. Skjeseth (1963) erected the "Heramb Shale and

Fig. 14. Detail of the upper part of the Herram section showing the intercalating limestone beds and shales of the Herram Member.

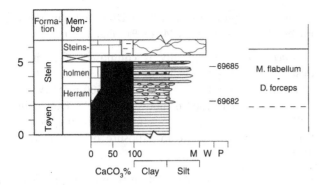

Fig. 15. Stratigraphic column for the Herram section. Conodont zones are indicated to the right. See Fig. 8 for legend.

Limestone" as a part of the Tøyen Formation ("Lower Didymograptus Shale") and the overlying "Stein Limestone" as an equivalent to the "Orthoceras Limestone" (Huk Formation) of the Oslo Region.

Owen *et al.* (1990) revised the "Herram Shale and Limestone" and referred the unit to the Herram Member and the "Stein Limestone" to the Stein Member of the Huk Formation. Subsequent investigations, however, have shown that the Lower Allochthon "Stein Limestone" and the "Heramb Shale and Limestone" are significantly different from the Huk Formation of the Oslo Region. Consequently, Rasmussen & Bruton (1994) erected the Stein Formation and included the Herram Member as the lower member and renamed the "Stein Limestone" the Steinsholmen Member and included it as the upper member of the Stein Formation.

Stratigraphic succession. – About 1 m of dark grey, graptolitic shales of the Tøyen Formation is exposed in the lower part of the section.

The overlying Herram Member of the Stein Formation is composed of grey shale with 8 horizons characterised by scattered limestone nodules. The thickness approximates to 1 m.

The Herram Member is succeeded by the Steinsholmen Member. The base is defined by the first coherent limestone bed. The lower c. 1.5 m is composed of grey shale with 14 nodular and coherent limestone beds. The coherent beds displays the reticulate pattern that characterises the overlying massive part of the Steinsholmen Member. The limestones of the Herram Member and the Steinsholmen Member contain shelly fossils and may be classified as mud- and wackestone.

Conodont biostratigraphy. – All of the Herram Member and the lower part of the Steinsholmen Member, composed of alternating shale and limestone beds, correlate with the *M. flabellum – D. forceps* Zone. This is in accordance with the graptolite dating, as *D. hirundo* Zone graptolites have been discovered in the

Fig. 16. Detail of the Elvdal Formation at Høyberget.

Herram Member (Kristina Lindholm, pers. comm. 1992).

Høyberget, Norway

Location. – The locality crops out c. 4 km southeast of Elvdal and 1 km east of Snerta (map sheet Elvdal 2018 III, UTM PP452441) (Fig. 1: 9).

Characteristics. – Limestones of the Lower Allochthon Elvdal Formation are situated on the western slope of Høyberget, mainly as erratic blocks. The studied samples from the Elvdal Formation are from a poorly exposed section, and it was not possible to determine the exact thickness because of faulting and overburden (Fig. 16). At the stratotype locality c. 1 km to the south of this locality (Rasmussen & Bruton 1994), at least 5 – 6 m of the upper part of the Elvdal Formation crops out in the small stream to the northeast above the road. Below the road 1 metre crops out just above the upper Proterozoic Vardal Sandstone of the Vangsås Formation. The estimated thickness at this site is about 14 m. The dip at the stratotype locality at Høyberget varies between 170°/22°E and 0°/15°E.

Previous work. – Fossils from the Elvdal Formation were

listed by Kjerulf 1863; nautiloid cephalopods and gastropods), Schiøtz (1874; trilobites), Henningsmoen (1979; brachiopods, ostracods and echinoderms) and Spjeldnæs (1985; cephalopods). The latter author also included a discussion of the stratigraphic position of the limestones at Høyberget.

The Høyberget area was described by Meinich (1881), Schiøtz (1883), Bjørlykke (1905) and mapped by Holtedahl (1921) and Nystuen (1975). Holtedahl (1921) estimated the thickness of the "Orthocerkalk" (=Elvdal Formation) to 5–10 m.

Rasmussen & Stouge (1989) studied the conodont fauna and showed that the limestone at Høyberget is considerably younger than the Lower Ordovician Huk Formation (previously named "Orthoceras Limestone") and instead is upper Llanvirn–early Caradoc in age.

The limestone at Høyberget and equivalent localities were formally described and revised by Rasmussen & Bruton (1994) who named the unit the Elvdal Formation.

Fossils from the succeeding black shale include asaphid trilobites (Holtedahl 1921), brachiopods (Henningsmoen 1979), cephalopods (Holtedahl 1921), echinoderms (Schiøtz 1874; Bjørlykke 1905), and gastropods (Schiøtz 1874; Bjørlykke 1905; Holtedahl 1921). Up till now precise correlation of this unit has not been possible on the basis of the macrofauna due to the poor preservation.

Stratigraphic succession. – The Elvdal Formation consists of medium grey, poorly-bedded, biogenic, argillaceous lime mudstone with a characteristic irregular, reticulate weathered surface. The Elvdal Formation is succeeded by a black, poorly fossiliferous, Ordovician shale unit which is about 20 m thick at Høyberget (Bjørlykke 1905). The shale unit is nodular in the lowermost part. The Elvdal Formation and the succeeding black, poorly fossiliferous, Ordovician shale unit rest unconformably on Vendian siliciclastic sediments of the Lower Allochthon Vangsås Formation. Middle Allochthon, Vendian, siliciclastic sediments referred to the Høyberg Formation of the Kvitvola Nappe is thrust onto the Lower Allochthon sediments.

Conodont biostratigraphy. – The fauna includes *Pygodus anserinus* Lamont & Lindström and *Baltoniodus variabilis* (Bergström) which correlates with the upper part of the *P. anserinus* and the lower part of the *A. tvaerensis* conodont zones of Bergström (1971), indicative of a latest Llanvirn to Early Caradoc age.

Engerdal, Røskdalsknappen, Norway

Location. – The locality is situated in a small valley on the northern flank of the mountain Røskdalsknappen, about 9 km south of the village Drevsjø (map sheet Engerdal 2018 I, UTM UJ424569) (Fig. 1: 10).

Characteristics. – The locality is composed of numerous, large limestone blocks of the Elvdal Formation situated in the small valley about 200–500 m NNW of Røskdalsknappen (1000 m above sea level). Holtedahl (1921, p. 38) mentioned a dip towards the north. Sample 97655 was from an exposure orientated 18°/36°NW.

Previous work. – The Engerdal area was mapped by Holtedahl (1921) and Nystuen (1974), and the stratigraphy was discussed by Rasmussen & Bruton (1994). Both Holtedahl and Nystuen correlated the limestone at Røskdalsknappen with the "Orthoceras Limestone" (=the Huk Formation of the Oslo Region), but the conodont data demonstrate that it is considerably younger and belongs to the Elvdal Formation. Holtedahl (1920) noted that the limestone at Røskdalsknappen is identical with the better known exposures at Høyberget.

Stratigraphic succession. – The Elvdal Formation consists of grey, weakly-bedded, biogenic, argillaceous lime mudstone with a characteristic irregular, reticulate weathered surface, similar to that of the Stein Formation. The vertical distance between the uppermost and the lowermost limestone blocks of the Elvdal Formation is about 30 m along the small valley at the northern slope. The uppermost block was obtained from altitude 940 m and the lowermost from 910 m, but it is impossible to estimate the accurate thickness. One significant exposure at an altitude of 925 m (about 5 × 2 × 1 m) was possibly *"in situ"* but erratic blocks dominate throughout the small valley. The limestone is succeeded by dark shales at an altitude of 950 m.

Conodont biostratigraphy. – The two investigated samples (97655, 97656) contained *Pygodus anserinus*, *Scabbardella altipes* and *Eoplacognathus* sp. and correlate with the *Pygodus anserinus* Zone (latest Llanvirn–Early Caradoc).

Sorken, Norway

Location. – 2 km northwest of the small village Sorken and 12 km north of Drevsjø, close to the southern end of Lake Femunden (map sheet Engerdal 2018 I, UTM PP562768) (Fig. 1: 11).

Characteristics. – The locality consists of large limestone blocks of the Elvdal Formation situated on both sides of the road at Langtjern. Only loose blocks were observed. The spot is completely surrounded by

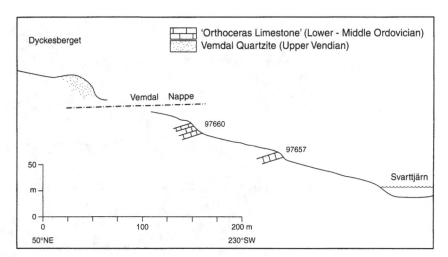

Fig. 17. Schematic cross section of the locality Glöte, Dyckesberget. The autochthonous, Ordovician "Orthoceras Limestone" is succeeded by allochthonous, Vendian sandstones and conglomerates of the Vemdal nappe. The position of the conodont samples is indicated.

Vendian siliciclastic sediments of the Vangsås and Ring formations.

Previous work. – The area was mapped by Holmsen (1937) and Nystuen (1974).

Stratigraphic succession. – The Elvdal Formation consists of grey, reticulate lime-mudstone. Some of the erratic blocks are several metres in diameter, but it is impossible to estimate the thickness of the succession.

Conodont biostratigraphy. – The two examined samples contained a very sparse and badly preserved fauna including *Protopanderodus robustus*, "*Drepanoistodus venustus*" and *Baltoniodus* sp. The three taxa co-occur in the interval from Llanvirn to Caradoc and are of little stratigraphical importance.

Glöte, Sweden

Location. – C. 2 km northwest from the village of Glöte, Härjedalen, on the southwestern slopes of the mountain Dyckesberget (UTM 13772E 68958N).

Characteristics. – Limestones are sporadically exposed in the forest around Dyckesberget. The present study concentrated on the area between the northern end of Svarttjärn and the northwestern part of Dyckesberget (Fig. 17). The limestone dips towards the north (commonly about 90°/37°N), but the dip may vary between 63°/38°NNW and 142°/45°NE.

Previous work. – The stratigraphical and tectonical position of the limestone succession ("Ortocerkalk") at Dyckesberget and other localities in relation to the surrounding siliciclastic deposits (previously "sparagmites") were discussed by Schiøtz (1883, 1892) and Högbom (1891, 1920). Schiøtz argued that the Vemdal Quartzite was placed on top of the limestone by

inversion, while Högbom was of the opinion that the Vemdal Quartzite simply was deposited later than the limestone and thus was of Post-Ordovician ("Undra Silur") age. Törnebohm (1896) introduced the controversial hypothesis that the Vemdal Quartzite partly was thrust onto the limestone. It has been suggested that the Vendian Vemdal Quartzite correlates with the Vangsås Formation of southeastern Norway, and is part of the Lower Allochthon (Asklund 1933; Röshoff 1978; Nystuen 1981), whereas the underlying limestone is autochthonous.

Stratigraphic succession. – The total thickness of the limestone sequence approximates to 40 m, but it is difficult to estimate due to the sporadic exposures at this locality. The lowermost exposure (sample 97657) occurs about 615 m above sea level, which is 20 m above Lake Svarttjärn. The limestone is grey and well-bedded, with about 10 cm thick beds. Thin intercalating shale beds occur sporadically within the limestone. The most continuous section observed is 7 m thick and is situated about 628 – 635 m above sea level. The stratigraphically youngest sample (97660) was collected from a 3 m thick section at an altitude of 635 m about 30 m NNW of the 7 m thick section.

The lithology and the conodont content indicate that the succession is autochthonous and belongs to the Central Baltoscandian Confacies Belt (Jaanusson 1976), and probably includes the interval from the Holen Limestone to the Furudal Limestone.

Lower Allochthonous, Vendian siliciclastic sandstones (Vemdal Quartzite) were observed at an altitude of 660 – 680 m. The interval between 635 m and 660 m was covered by drift.

According to Högbom (1920, fig. 43) alum shale occurs below the limestone close to the shore of Svarttjärn. No shales were discovered during the present study, but this might be due to overburden.

Fig. 18. The upper part of the Stein Formation at Andersön-A (view from west).

Conodont biostratigraphy. – The limestone correlates with the interval from the *Lenodus variabilis* or possibly *E. suecicus* Zone to the *Pygodus anserinus* Zone of the Baltoscandian standard conodont zonation scheme.

Andersön-A, Sweden

Location. – Map sheet Östersund 193B, UTM 14307E 70074N. The Andersön-A section is situated on the north coast of the island Andersön some 16 km west of Östersund. The locality is a nature preservation area.

Characteristics. – The upper c. 10 m of the Tøyen Formation and the lower 17 m of the Stein Formation are well exposed in the shore section. The interval from 17 to c. 29 m above the base of the Stein Formation was covered by soil and plants, but it was possible to uncover small exposures of the succession by digging. The section dips towards the west (on average 178°/58°W).

Previous work. – The Andersön-A section was studied by Hadding (1912, locality 8), Tjernvik (1956) and Bergström (1988). Hadding (1912) described the section briefly and mentioned the presence of *Phyllograptus* sp. and *Didymograptus* about 2.5 m below the top of the Tøyen Formation.

A detailed biostratigraphy of the Tøyen Formation was given by Tjernvik (1956) based on graptolites and trilobites. Tjernvik (1956, pp. 171 – 175) correlated the succession with the *M. armata, P. planilimbata / Tetragraptus phyllograptoides* and *D. balticus* zones, and

reported *Megistaspis* sp. from the overlying "Limbata limestone" (= Stein Formation). Although the locality has been studied carefully by the present author, it has not been possible to recognise the 0.6 m thick limestone bed that was correlated with the *M. armata* zone by Tjernvik.

Bergström (1988) presented a detailed description of the *Prioniodus elegans* fauna from the Andersön-A section and other localities.

Stratigraphic succession. – The exposed section is about 40 m thick. The lower part consists of dark, graptolitic shales of the Tøyen Formation, c. 10 m in thickness. Limestone lenses and thin limestone beds are common in the upper 4 m.

The upper part comprises about 29 m of the Stein Formation (Figs. 18, 19). All but the 3 m thick interval 18 – 21 m above the base of the formation, have been referred to Lithofacies A, which is characterised by micritic, grey, mainly reticulate mud- and wackestone (Rasmussen & Bruton 1994). The remaining part consists of a more pure wackestone which have been referred to Lithofacies C.

Conodont biostratigraphy. – The upper 4 m of the Tøyen Formation contain conodont faunas indicative of the interval between the *P. proteus* Zone and the *M. flabellum – D. forceps* Zone. The Stein Formation correlates with the interval between the *M. flabellum – D. forceps* Zone and the *M. ozarkodella* Zone.

Andersön-B, Sweden

Location. – The island of Andersön, map sheet Öster-

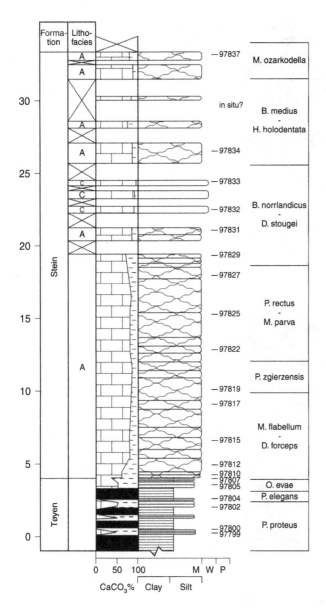

Fig. 19. Stratigraphic column of the Andersön-A section. Conodont zones are indicated to the right. See Fig. 8 for legend.

sund 193B, UTM 14297E 70066N. The section is exposed just north of the bay on the west coast of Andersön. The locality is a nature preservation area.

Characteristics. – The Andersön-B section is a shore section. The lower c. 20 m of the section is visible on the southern shore close to the water surface, while the upper about 25 m forms 2–5 m high cliffs on the western shore (Fig. 20A–C). All parts are easily accessible. The section is faulted, and the upper part of the Stein Formation is tectonically repeated.

The lowermost samples correlate biostratigraphically with the Tøyen Formation of the Andersön-A section, but only very thin shale intervals are visible at Andersön-B.

The section dips towards the northwest, varying between 33°/29°NW in the lower part through 16°/42°NW in the middle part to 38°/43°NW in the upper part.

Previous work. – The section has earlier been studied by Linnarsson (1872), Wiman (1893, 1898), Hadding (1912, locality 1), Bergström (1971), Bergström *et al.* (1974), Karis (1982). Linnarsson (1872) gave a lithologic description, and a list of fossils, mainly trilobites, from the "Biseriata Limestone" (= "Chasmopskalk" of Linnarsson).

The presence of *Ogygiaschiefer* (= "Andersö Shale") was mentioned by Wiman (1893). Five years later, he (Wiman 1898, p. 274) presented an illustration of the Andersö-B section including the interval from the upper part of the Stein Formation to the dark shales overlying the "Biseriata Limestone".

Hadding (1912, p. 599) recorded *Diplograptus putillus* from shales overlying the top of the Stein Formation. Based on the findings, he correlated the interval with the Lower "Dicellograptus Shale" in Scania (*H. teretiusculus* Zone).

The conodont biostratigraphy of the "Biseriata Limestone" was discussed by Bergström (1971, locality 26) and further demonstrated by Bergström *et al.* (1974). The latter authors showed that *Pygodus serra* and *P. anserinus* co-occur within the "Biseriata Limestone". The present work failed to confirm the presence of *P. serra* from the "Biseriata Limestone", although it was detected from limestone nodules about 2 m below the base.

The allochthonous Ordovician succession of Jämtland was described and discussed by Karis (1982). He introduced the informal names Isö Limestone for the Stein Formation and Andersö Shale for the succeeding shale interval, previously named the Ogygiocaris Shale. Subsequent studies, however, revealed that the Isö Limestone is identical with the Norwegian "Stein Limestone" (Skjeseth 1963) and both were included within the Stein Formation by Rasmussen & Bruton (1994). Karis (1982) noted the occurrence of *Asaphus expansus* about 5 m above the base of the section, and *Paraceraurus excul* (Beyrich) and *Pseudomegalaspis patagiata* (Törnquist) in the uppermost 2 metres. The latter two species indicate correlation with the Folkslunda Limestone according to Karis (1982).

Stratigraphic succession. – The lower about 29 m consists of micritic mud- and wackestone of the Stein Formation (Fig. 21). However, it is possible that the lowermost 1–1.5 m may be part of the Tøyen Formation (see above). The interval from 10 to 13 m above the base consists of relatively pure wackestone of Lithofacies C (Rasmussen & Bruton 1994), while most of the remaining part is typical

(A)

(B)

(C)

Fig. 20. The Andersön-B section. □A: The middle part of the Stein Formation (view from west). □B, C: The upper part of the Stein Formation and the lower part of the succeeding "Andersö Shale".

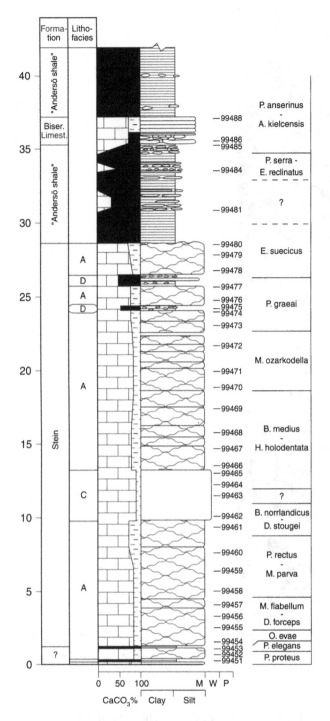

Fig. 21. Stratigraphic column of the Andersön-B section. Conodont zones are indicated to the right. See Fig. 8 for legend.

developed at Andersön-B, whereas the lower part is most complete at Andersön-A. Together, the two sections indicate a total thickness of about 38 m of the Stein Formation, which is in accordance with the estimate of Karis (1982; 40 m), but is considerably thicker than noted by Hadding (1912; 15 m).

The Stein Formation is overlain by about 7 m dark faulted and folded shales of the "Andersö Shale" that is nodular in the upper 5 m. The shale interval is succeeded by a 1.5 m thick limestone unit, the "Biseriata Limestone", which probably is the eastern equivalent of the Elvdal Formation (Fig. 22A, B). Dark shales of the upper Andersö Shale of unknown thickness overlie the unit.

Conodont biostratigraphy. – The Tøyen Formation and Stein Formation span the interval from the *P. proteus* Zone to the *E. suecicus* Zone. A limestone nodule from the overlying shale interval is indicative of the *P. serra – E. reclinatus* Subzone, while the "Biseriata Limestone" correlates with the *P. anserinus-S. kielcensis* Subzone.

Andersön-C, Sweden

Location. – 30 m west of the Andersön-A section. Map sheet Östersund 193B, UTM 14307E 70074N. The locality is a nature preservation area.

Characteristics. – The section is an incomplete shore section which is largely covered by drift. The beds dip towards the northwest (about 15°/37°NW).

Previous work. – The locality was visited by Hadding (1912), who indicated the presence of "Orthocerkalk" (= Stein Formation) on the geological map just northwest of his locality 8 (Hadding 1912).

Stratigraphic succession. – The two investigated spot samples indicate that the interval between the uppermost part of the Stein Formation and the "Biseriata Limestone" is present. The section is dominated by nodular shales.

Conodont biostratigraphy. – Interval between the *Protopanderodus graeai* Zone and the *Pygodus anserinus* Zone.

Technique

The conodont elements have been recovered by standard procedures (Lindström 1964; Stouge & Boyce 1983).

The conodont elements from the 133 processed limestone samples were isolated using 12% acetic acid. Each sample weighed 1.5 kg except the samples from

---

Lithofacies A, composed of reticulate, grey mud- and wackestone. Two thin intervals about 24 – 27 m above the base comprise thin nodular horizons surrounded by shales, which were referred to Lithofacies D. Large orthoconic cephalopods characterise the upper 10 m of the Stein Formation.

The upper part of the Stein Formation is well

(A)

(B)

Fig. 22. A, B. The "Biseriata Limestone" and the upper, nodular part of the preceding "Andersö Shale". Andersön-B. The beds are dipping towards the lake.

Herram and Steinsodden (0.7 kg) and Grøslii (on average 1.0 kg). The residue was washed through a 71 μm sieve and the material between 71 μm and 1 mm was separated in heavy liquid (specific gravity = 2.8 g/cm³). All of the heavy residue was picked.

The samples from the Stein Formation contained 164 conodont elements on average (160 elements/kg), whereas the samples from the Elvdal Formation yielded 10 specimens on average (= 7 elements/kg). 6 samples were barren. The figured specimens were photographed by Scanning Electron Microscope.

A total of 19,286 specimens were recovered, represent-

ing 398 morphotypes, which have been referred to 45 genera and 102 species. 206 specimens were unidentifiable.

Conodont stratigraphy

The Baltoscandian conodont zones established by Lindström (1971) and Bergström (1971, 1983) with modifications by Löfgren (1978, 1993a, b, 1994, 1995); Stouge (1989); Stouge & Bagnoli (1990) and Bagnoli & Stouge (1997) have been based primarily on sections from the Central Baltoscandian Confacies Belt (Jaanusson 1976). The Tremadoc – lower Arenig zones (*Paltodus deltifer, Paroistodus proteus, Prioniodus elegans* and *Oepikodus evae* zones) and middle – upper Llanvirn zones (*Eoplacognathus suecicus, Pygodus serra* and *Pygodus anserinus*) can be recognised and used without problems within the Lower Allochthon platform margin deposits studied here. However, more complex stratigraphic considerations are necessary for the late Arenig – early Llanvirn period, when the platform margin deposits and the inner platform deposits became dominated by quite different species associations. Additionally, facies-dependent taxa, for example *Lenodus* (including *L.* cf. *variabilis*), *Scalpellodus, Semiacontiodus, Protopanderodus* and *Periodon*, developed dissimilar stratigraphic ranges during this interval in response to the development of various local environmental conditions (Rasmussen & Stouge 1995). As a consequence seven new biozones are introduced for the critical interval: The *Microzarkodina flabellum – Drepanoistodus forceps* Concurrent Range Zone, *Periodon zgierzensis* Partial Range Zone, *Protopanderodus rectus – Microzarkodina parva* Interval Zone, *Baltoniodus norrlandicus – Drepanoistodus stougei* Concurrent Range Zone, *Baltoniodus medius – Histiodella holodentata* Oppel Zone, *Microzarkodina ozarkodella* Partial Range Zone and *Protopanderodus graeai* Interval Zone. The new zones are particularly useful for correlations between the Baltoscandian platform margin sections, but to improve the stratigraphic potential of the zones, they have been based on species allowing correlation with both the Baltoscandian standard zones (Lindström 1971; Bergström 1971) and the North American platform margin zones (Stouge 1984).

The zones are defined in accordance with the recommendations from the International Subcommission on Stratigraphic Classification of the IUGS Commission on Stratigraphy (Hedberg 1976) and the Norwegian Committee on Stratigraphy (Nystuen 1989).

It should be noted that the headings "*Characteristics*", "*Distribution and thickness*" and "*Occurrence*" refer only to sections studied here.

Paltodus deltifer Zone (Lindström 1971)

Original definition. – Conodont fauna dominated by *Cordylodus* Pander, *Oneotodus variabilis* (Lindström), *Paltodus deltifer* (Lindström), and *Paroistodus numarcuatus* (Lindström) (Lindström 1971, pp. 26, 27).

Characteristics. – The samples from Grøslii contained a very sparse conodont fauna dominated by *Paltodus* cf.

deltifer and *Paroistodus numarcuatus*. "*Scolopodus*" *peselephantis* is also present.

Distribution and thickness. – The Solheim Slate Member of the Ørnberget Formation at Grøslii Seter, Valdres: unknown thickness; Røste 0.3 m.

Remarks. – According to the original definition (Lindström 1971), the *P. deltifer* Zone may probably be classified as an assemblage zone. Only few specimens were obtained from this zone. The Røste sample was collected from nodular horizon about 9 m below the base the Stein Formation. It contained only a few identifiable specimens including *Paltodus deltifer, Paltodus* cf. *subaequalis* and *Paroistodus numarcuatus* (Fig. 23, Appendix 5).

Paroistodus proteus Interval Zone (Lindström 1971 emended herein)

Original definition. – Lindström (1971, p. 29) notes that *Paroistodus proteus* and *Paltodus inconstans* are the two leading species of the zone.

Emended definition. – Interval from the first occurrence of *Paroistodus proteus* to the first occurrence of *Prioniodus elegans*.

Characteristics. – The interval is dominated by *Paroistodus numarcuatus, Paltodus* cf. *subaequalis* and *Paroistodus proteus*, whereas *Drepanodus arcuatus, Scolopodus peselephantis*, Gen. *et* sp. indet. A and *Drepanoistodus* sp. indet. are relatively common within the Bjørkåsholmen Formation at Steinsodden. *Cornuodus longibasis* occurs sparsely. The high number of *Paroistodus numarcuatus* together with the occurrence of *Paltodus* cf. *subaequalis* is typical of the horizon close to the boundary between the *P. deltifer* and *P. proteus* zones.

The upper sample from the Bjørkåsholmen Formation at Steinsodden contained a few specimens representing *Paroistodus proteus, Paltodus* cf. *subaequalis, Cornuodus longibasis*, and *Drepanoistodus* sp. which is a typical *P. proteus* Zone fauna.

The basal sample within the Andersön-B section (sample 99451) is dominated by *Paracordylodus gracilis, Paroistodus proteus* and *Tetraprioniodus robustus* (Fig. 24, Appendix 3). Few specimens of *Paltodus subaequalis* and *Scolopodus quadratus* were recorded. The overlying sample (99452) yields a similar fauna, except that *Oelandodus elongatus, Drepanoistodus forceps* and *Drepanodus arcuatus* only are sparsely represented.

Paracordylodus gracilis also dominates the two lowermost samples at the Andersön-A section (sample 97799 and 97800). *Paroistodus proteus*,

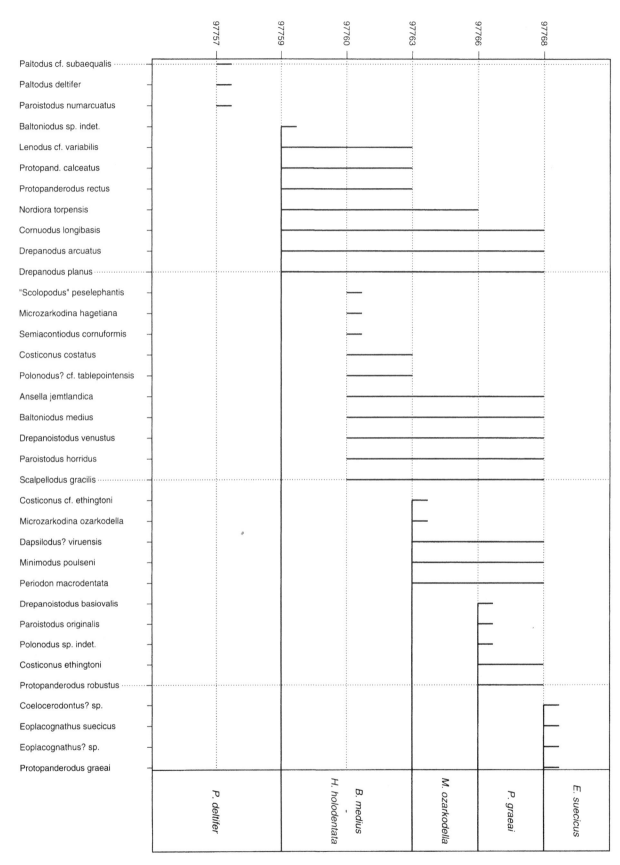

Fig. 23.   Range chart of conodonts from the Røste section.

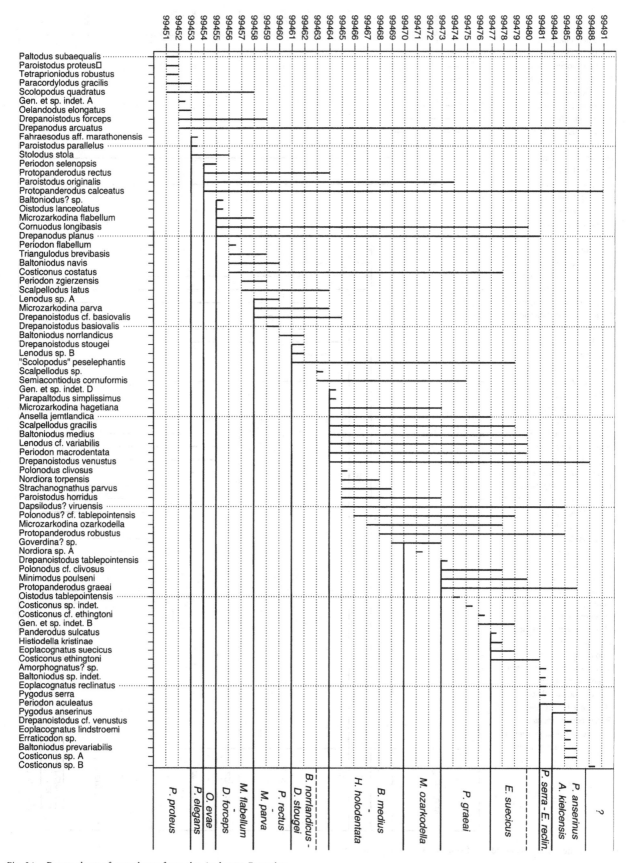

Fig. 24. Range chart of conodonts from the Andersön-B section.

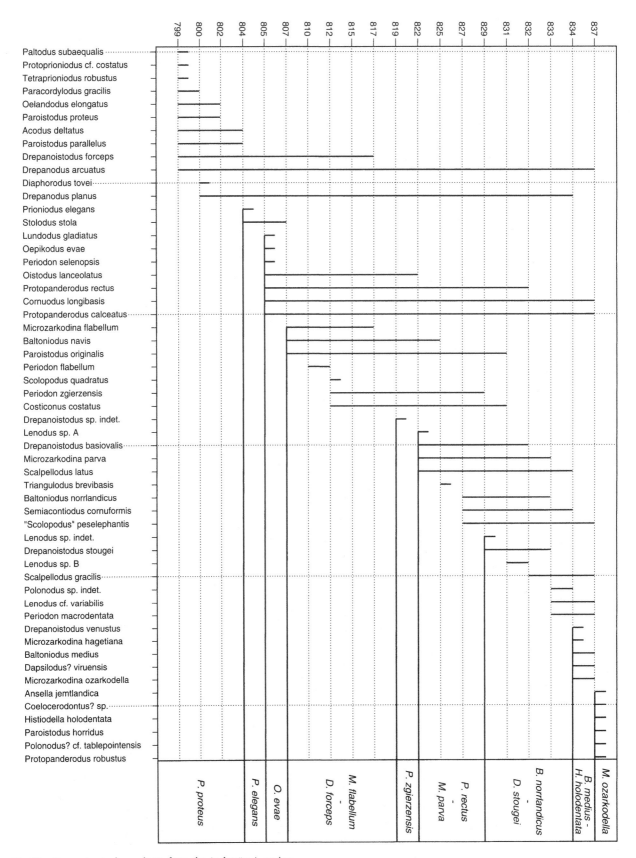

Fig. 25.   Range chart of conodonts from the Andersön-A section.

Oelandodus elongatus, Paroistodus parallelus and *Drepanoistodus forceps* are relatively common. *Tetraprioniodus robustus* and *Paltodus subaequalis* occur in sample 97799 but not in the overlying 97800 (Fig. 25, Appendix 2). Sample 97802 is from a limestone bed 1.55 m above 97800 at the Andersön-A section. It contained a very sparse fauna dominated by *Oelandodus elongatus, Paroistodus. parallelus* and *Drepanoistodus forceps*. One *Stolodus stola* specimen was recorded.

Distribution and thickness. – Steinsodden + 0.7 m; Andersön-A + 1.9 m; Andersön-B + 0.4 m.

Prioniodus elegans Interval Zone (Lindström 1971, emended Stouge & Bagnoli 1988)

Definition. – The interval between the first occurrence of *Prioniodus elegans* and the first occurrence of *Oepikodus evae* (Stouge & Bagnoli 1988).

Characteristics. – Sample 97804 from the Andersön-A section yielded only 26 specimens. This means that the calculated relative abundances must be handled with care. The sample is dominated by *Drepanoistodus forceps, Prioniodus elegans* and *Paroistodus proteus*, whereas *Drepanodus planus, Oelandodus elongatus* and *Stolodus stola* occur sporadically.

Only 13 specimens were recorded from sample 99453 from the Andersön-B section. The sample did not contain *P. elegans*, but the presence of *Fahraeusodus* sp., *Oelandodus elongatus, Paracordylodus gracilis, Paroistodus parallelus* and *Stolodus stola* corresponds with the *P. elegans* Zone fauna from Kalkberget, Jämtland (Löfgren 1978), Diabasbrottet, Västergötland (Löfgren 1993a) and Talubäcken, Siljan (Bergström 1988).

Distribution and thickness. – Andersön-A: + 0.2 m; Andersön-B: 0.6 m.

Remarks. – *Prioniodus elegans* Zone conodonts were recovered from the Andersön-A section by Bergström (1988, p. 235, text-fig. 7). He gave no details of the fauna but reported that "an excellent "pure" *P. elegans* fauna has been isolated from a sample collected 30 cm above the base of Tjernvik's (1956, pp. 171–173) Unit E..". Bergström's sample is undoubtedly from the same bed as the one described here. The *P. elegans* bearing interval is generally very thin on the Baltoscandian platform and is often preserved only as a reworked "ghost fauna" in the bottom of the *Oepikodus evae* Zone (Lindström 1971). However, the following

Baltoscandian *P. elegans* Zone localities have been described in recent years: Leningrad, Russia: 0.40 m (Bergström 1988), Talubäcken, Dalarna, Sweden: 0.55 m (Bergström 1988), Lanna, Närke, Sweden: 0.10–0.15 m (Bergström 1988), Äleklinta, Öland, Sweden: 0.40 m (van Wamel 1974), Köpingsklint, Öland, Sweden: Less than 0.20 m (Bagnoli *et al.* 1988). A presumed more than 8 metres thick sequence (mainly shales) correlated with the *P. elegans* Zone was described from Diabasbrottet, Västergötland, Sweden by Löfgren (1993a, fig. 7).

Oepikodus evae Interval Zone (Lindström 1971, emended herein)

Original definition. – Lindström (1971, pp. 29–30) describes the *Oepikodus evae* Zone as the interval between the first apperance of *Oepikodus evae* and the first appearance *Baltoniodus triangularis*.

Emended definition. – Interval between the appearance of *Oepikodus evae* (Lindström) and the appearance of *Microzarkodina flabellum* (Lindström).

Characteristics. – Only one sample (97805) from the Andersön-A section is correlated with the *Oepikodus evae* Zone. The sample is dominated by *Drepanoistodus forceps* and *Oepikodus evae*. *Periodon selenopsis* and *Stolodus stola* are also characteristic, while *Protopanderodus rectus, Protopanderodus calceatus, Paroistodus parallelus* and *Lundodus gladiatus* are less common. *Oistodus lanceolatus, Drepanodus planus* and *Cornuodus longibasis* occur sporadically.

Sample 99454 from the Andersön-B section contained only few specimens, which were dominated by *Drepanoistodus forceps*, while *Periodon selenopsis, Paroistodus originalis, Drepanodus arcuatus, Protopanderodus rectus* and *P. calceatus* occurred in low numbers.

Distribution and thickness. – Andersön-A: 0.2 m; Andersön-B: 1.0 m.

Remarks. – The *B. triangularis* Zone has not been observed in the present study. Consequently, the upper boundary of the *O. evae* Zone has been placed at the base of the *Microzarkodina flabellum – Drepanoistodus forceps* Zone. In Sweden, *Oepikodus evae* is missing in the upper part of the *O. evae* Zone (Lindström 1971; Löfgren 1978, 1985, 1993b), whereas it was recorded from the *M. flabellum – D. forceps* Zone from the locality Herram within the Lower Allochthon of southeastern Norway. It is thus apparent that the vertical range of *Oepikodus evae* is strongly related to the different palaeoenvironments on the Baltoscandian platform. Stouge & Bagnoli (1988) introduced the *Oepikodus evae* Range Zone for the slope deposits of the Cow Head Group, western Newfoundland.

Microzarkodina flabellum – Drepanoistodus forceps Concurrent Range Zone (new zone)

Definition. – Interval where *Microzarkodina flabellum* and *Drepanoistodus forceps* have concurrent ranges.

Reference section. – The Andersön-A section (UTM 14307E 70074N), Andersön, Jämtland, Sweden.

Range. – The uppermost 30 cm of the Tøyen Formation and the lower 5.7 m of the Stein Formation.

Characteristics. – The lower part of the *M. flabellum – D. forceps* Zone is characterised by *Drepanoistodus forceps* as the dominant species. Other common species in the lower part of the Zone at Andersön are *Protopanderodus rectus* and *Microzarkodina flabellum*. *Periodon selenopsis* occurs in basal part of the zone at Andersön-B, while it has been recorded from the *O. evae* Zone at Andersön-A. It is possible that the relatively late occurence of *P. selenopsis* at Andersön-B is caused by reworking, but no clear signs of that have been observed. The middle and upper parts of the zone are dominated by *Drepanoistodus forceps*, *Protopanderodus rectus* and *Paroistodus originalis*. Sample 99456 from the middle part of the *M. flabellum – D. forceps* Zone at Andersön-B is dominated by *Microzarkodina flabellum*. *Periodon flabellum*, *Costiconus costatus*, *Baltoniodus navis* and *Drepanodus* spp. occur sparsely in the middle and upper parts of the zone at Andersön, while *Triangulodus brevibasis* has been recorded from the zone only within the Andersön-B section.

The Herram section and the very basal part of the Steinsholmen section have a rather unusual species composition (Figs. 12, 15, 26). It was noted already by Kohut (1972, p. 437) that the sections comprise a mixture of "older species" and "younger species". Members of the first group include *Paracordylodus gracilis*, *Stolodus stola* and *Oepikodus evae*, while *Trapezognathus quadrangulum*, *Baltoniodus navis*, *Microzarkodina flabellum*, *Periodon flabellum*, *Periodon zgierzensis* and *Paroistodus originalis* are typical of the second group. The lithology is dominated by shales with limestone nodules and thin, coherent limestone beds (Skjeseth 1963; Owen *et al.* 1990).

Drepanoistodus forceps, *Microzarkodina flabellum* and *Periodon flabellum* dominate the lower sample at Herram (69682) (Fig. 26, Appendix 4) whereas *Protopanderodus rectus* and *O. evae* are common. *Trapezognathus quadrangulum*, *Stolodus stola*, *Paracordylodus gracilis* and *Baltoniodus navis* occur sparsely. The sample from the middle part of the Herram section (69685) is distinguished from the lower sample by a markedly decrease in *D. forceps*. In contrast, *P. originalis* becomes the most frequent species. *P. rectus*, *P. flabellum*, *B. navis* and *O. evae* are common, whereas *P. gracilis* and *S. stola* occur in low numbers. *Periodon zgierzensis* appears for the first time.

The composition of the lowermost sample at Steinsodden (69687) is quite similar to that of sample 69685. The most remarkable changes are the decrease in the relative abundance of *B. navis* and *P. flabellum*, and the disappearance of *Paracordylodus gracilis*. *Drepanoistodus* cf. *basiovalis* appears for the first time.

Distribution and thickness. – Andersön-A: 6.0 m; Andersön-B: 3.0 m; Herram: 2.8 m; Steinsodden: 0.2 m.

Remarks. – *Drepanoistodus forceps* ranges into the basal part of the *Microzarkodina parva* Interval Zone within the Andersön-B section.

Periodon zgierzensis Partial Range Zone (new zone)

Definition. – Interval between the last occurrence of *Drepanoistodus forceps* (Lindström) and the first occurrence of *Microzarkodina parva* Lindström.

Characteristics. – The *Periodon zgierzensis* Zone is dominated by *Paroistodus originalis* (Sergeeva) and *Protopanderodus rectus* (Lindström) at Steinsodden, whereas *Periodon zgierzensis* is abundant in the middle part (sample 69606). Other relatively common species include *Baltoniodus navis*, *Drepanoistodus* cf. *basiovalis* and *Drepanodus arcuatus*.

Reference section. – The Steinsodden section (UTM NN919539), Steinsodden, Moelv, Norway.

Range. – 0.2 – 2.6 m above the base of the Steinsholmen Member.

Distribution and thickness. – Steinsodden: 2.4 m; Andersön-A: 2.4 m.

Remarks. – The top of the zone coincides with the last occurrence of *Drepanoistodus* cf. *basiovalis* and *Triangulodus amabilis* at Steinsodden. *Periodon zgierzensis* is not restricted to the *P. zgierzensis* Zone, but ranges from the *M. flabellum – D. forceps* Zone (Arenig) to the *B. medius – H. holodentata* Zone (Llanvirn).

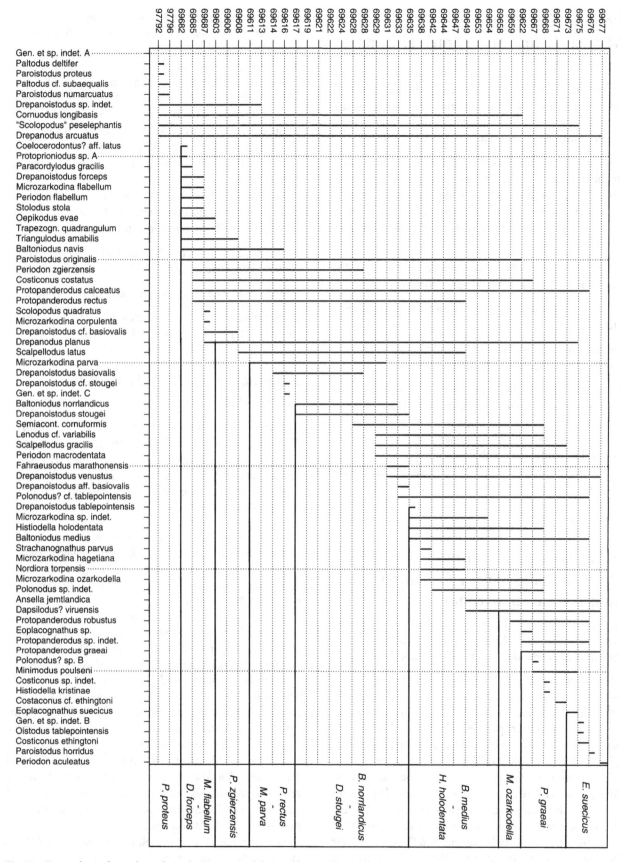

Fig. 26. Range chart of conodonts from the Herram and Steinsodden sections.

Protopanderodus rectus – Microzarkodina parva Interval Zone (new zone)

Definition. – Interval between the appearance of *Microzarkodina parva* Lindström and the level where *Baltoniodus norrlandicus* Löfgren and *Drepanoistodus stougei* Rasmussen co-occur for the first time.

Characteristics. – *Paroistodus originalis* is the most frequent species in the *Protopanderodus rectus – Microzarkodina parva* Zone. Other common species include *Baltoniodus navis*, *Protopanderodus rectus*, *Drepanodus arcuatus*, *Periodon zgierzensis* and *Microzarkodina parva*.

Reference section. – The Steinsodden section (UTM NN919539), Steinsodden, Moelv, Norway.

Range. – 2.7–8.0 m above the base of the Steinsholmen Member.

Distribution and thickness. – Steinsodden: 5.3 m; Andersön-A: 6.9 m, Andersön-B: 3.5 m.

Remarks. – Despite the fact that the base of the *Protopanderodus rectus – Microzarkodina parva* Zone is defined solely on the first occurrence of *M. parva*, it is chosen to introduce the combined name for the zone to avoid confusion with the completely younger *Microzarkodina parva* Zone established by Lindström (1971).

Baltoniodus norrlandicus – Drepanoistodus stougei Concurrent Range Zone (new zone)

Definition. – The *Baltoniodus norrlandicus – Drepanoistodus stougei* Zone is defined by the co-occurrence of *Baltoniodus norrlandicus* and *Drepanoistodus stougei*.

Characteristics. – The first overlap of *Baltoniodus norrlandicus* and *Drepanoistodus stougei* appears in sample 69617 at the reference section. The most numerous species in the lower part of the *B. norrlandicus – D. stougei* Zone are *Paroistodus originalis*, *Periodon zgierzensis* and *B. norrlandicus*. Other common species are *D. stougei*, *Protopanderodus rectus* and *Microzarkodina parva*. The relative number of *Scalpellodus latus* specimens increases significantly compared with the underlying zone. At Andersön-A the first specimens of *B. norrlandicus* appears in sample 97827 and is by far the most numerous species. *Paroistodus originalis*, *Scalpellodus latus* and *Drepanoistodus basiovalis* are common.

In the middle and upper parts of the *B. norrlandicus – D. stougei* Zone *B. norrlandicus*, *Drepanoistodus stougei* and *Scalpellodus* spp. become the dominate taxa at Steinsodden and Andersön. *D. stougei* is especially abundant in the middle part, while *Scalpellodus latus* and/or *S. gracilis* dominates the upper part. The two genera are considered very sensitive to environmental changes (Rasmussen & Stouge 1995) and it is suggested that their relative abundances should not be used in regional correlations.

Protopanderodus rectus and *Periodon zgierzensis* may occasionally be common. *Lenodus* cf. *variabilis* appears very sporadically in the upper part of the zone.

Reference section. – The Steinsodden section (UTM NN919539), Steinsodden, Moelv, Norway.

Range. – 8.0 – 24.3 m above the base of the section.

Distribution and thickness. – Steinsodden: 16.3 m; Andersön-A: c. 11 m; Andersön-B: 3.3 m.

Baltoniodus medius – Histiodella holodentata Oppel Zone (new zone)

Definition. – The base of the *Baltoniodus medius – Histiodella holodentata* Zone is defined by the appearance of *Baltoniodus medius* (Dzik) and *Histiodella holodentata* (Ethington & Clark). The upper boundary is placed where *Microzarkodina ozarkodella* outnumbers *M. hagetiana*.

Characteristics. – The *B. medius – H. holodentata* Zone is characterised by the *B. medius*, *H. holodentata*, *N. torpensis*, *M. hagetiana* and *Strachanognathus parvus* assemblage. *Polonodus* has often its first appearance just below the base of the zone.

The reference section is characterised by a significant amount of *Protopanderodus rectus* through the lower and middle parts of the subzone. The basal part (sample 69635) is moreover distinguished by numerous *Cornuodus longibasis* and "*Drepanoistodus venustus*". *Protopanderodus calceatus* ranges through the zone and is fairly common. *N. torpensis* occurs sporadically in the Andersön and Stein sections, whereas it becomes more frequent in the deeper-water facies (Røste, Jøronlia) (Figs. 23, 27, Appendix 5). *Ansella jemtlandica* and *Paroistodus horridus* have not been observed in the reference section but occur in small numbers within the Andersön and Røste sections. *Protopanderodus robustus* may occur in the upper part of the zone.

The uppermost two samples at Steinsodden (69653 and 69654) contain very few specimens (31 and 5 respectively) but indicate that *B. medius* and *S. gracilis* are most frequent in this interval.

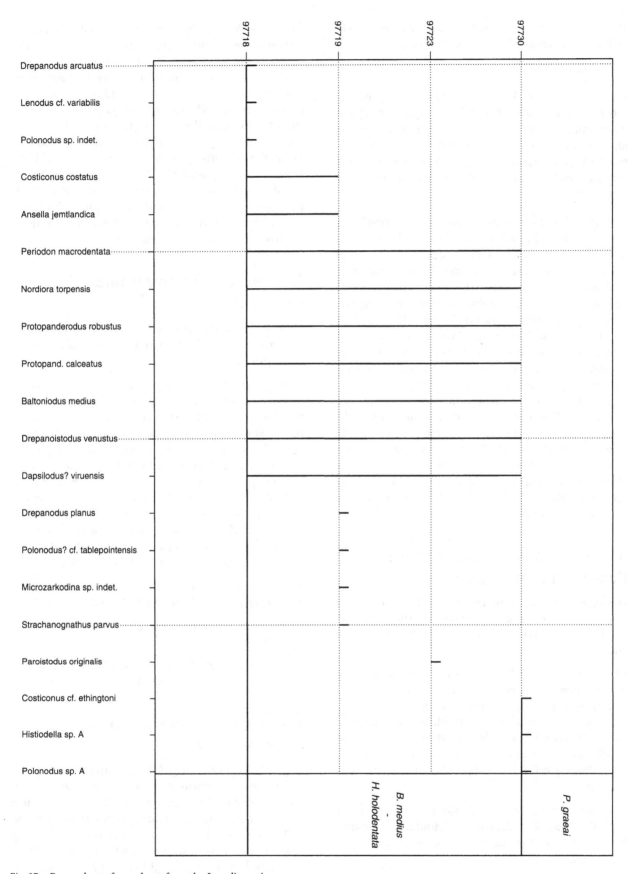

Fig. 27. Range chart of conodonts from the Jøronlia section.

Reference section. – The Steinsodden section (UTM NN919539), Steinsodden, Moelv, Norway.

Range. – 24.3 – 37.3 m.

Distribution and thickness. – Steinsodden: 13 m; Andersön-A: c. 7 m; Andersön-B: 6 m; Hestekinn: thickness unknown; Jøronlia: 5.2 m; Røste: 3 m. Faunas correlating with the *B. medius – H. holodentata* Zone have been recorded from spot samples also at the localities Haugnes and Skogheim. All samples are from the Stein Formation.

Remarks. – *Nordiora torpensis* n. sp., which is typical for the *B. medius – H. holodentata* Zone, is known only from the Scandinavian Caledonides, but species of *Nordiora* n. gen. have been reported from ocean-near, cold-water environments elsewhere. Dzik (1978) described *Phragmodus polonicus* from the *Pygodus serra – Eoplacognathus robustus* Subzone (Llanvirn) in the Holy Cross Mountains, Poland, and a related species is known from the Upper Arenig Komstad Formation, Bornholm, Denmark and Scania, Sweden (S. Stouge, pers. comm. 1993, in prep.).

Several species that occur within the *B. medius – H. holodentata* Zone have been recorded from outside the region, including the eastern part of the North American platform (see "Correlation"), and it was demonstrated by Rasmussen & Stouge (1988) that the Stein Formation contains faunas that allow a fairly precise correlation between the Baltoscandian and the eastern North American conodont zones.

Microzarkodina ozarkodella Partial Range Zone (new zone)

Definition. – Interval between the level where *Microzarkodina ozarkodella* outnumbers *Microzarkodina hagetiana* and the appearance of *Protopanderodus graeai*.

Characteristics. – The *M. ozarkodella* Zone is dominated by the nominal species together with *Scalpellodus gracilis* and *Baltoniodus medius* at Steinsodden. The faunal composition is somewhat different at Andersön where *Protopanderodus robustus*, *Microzarkodina ozarkodella* and *Periodon macrodentata* are dominant. *Scalpellodus gracilis* and *Baltoniodus medius* are relatively common.

Reference section. – The Andersön-B section (UTM 14297E 70066N), Andersön, Jämtland, Sweden. Range: 18.5 – 22.7 m above the base of the Stein Formation.

Distribution and thickness. – Andersön-B: 4.2 m; Andersön-A: +0.2 m; Steinsodden: 2 m; Røste: c. 2 m.

Remarks. – Within the Steinsodden and Andersön-B sections, few specimens of small *M. ozarkodella* may occur in the *B. medius – H. holodentata* Zone significantly below the level where *M. ozarkodella* becomes the most frequent *Microzarkodina*. In addition, these early occurrences of "*M. ozarkodella*" have been reported from the Finngrundet core by Löfgren (1985, p. 123), and do also occur in the upper Arenig deposits of Bornholm, Denmark (S. Stouge, pers. comm. 1990). Löfgren (1985) placed the lower boundary of the *E.? variabilis – M. ozarkodella* Subzone at the level where *M. ozarkodella* becomes the most numerous *Microzarkodina*, and that practice has been followed here to define the base of the *M. ozarkodella* Zone.

Protopanderodus graeai Interval Zone (new zone)

Definition. – Interval between the appearance of *Protopanderodus graeai* (Hamar) and *Eoplacognathus suecicus* (Bergström).

Characteristics. – The *P. graeai* Zone is dominated by *Protopanderodus graeai* and *Protopanderodus robustus* at both Steinsodden and Andersön. *Ansella jemtlandica* and *Baltoniodus medius* are relatively common in the lower part of the zone, while *Costiconus* cf. *ethingtoni* is common in the upper part. *Histiodella kristinae* appears in the upper part of the zone at Steinsodden, but at the base of the succeeding *E. suecicus* Zone at Andersön. *Minimodus poulseni* appears at the base of *Protopanderodus graeai* Zone at Andersön, and just above the base within the Steinsodden section. Farther west, at Røste, *M. poulseni* appears in the underlying *M. ozarkodella* Zone.

Reference section. – The Steinsodden section (UTM NN919539), Steinsodden, Moelv, Norway.

Range. – 40.5 – 44.0 m above the base of the Steinsholmen Member.

Distribution and thickness. – Steinsodden: 3.5 m; Andersön-B: 3.7 m; Jøronlia: +0.2 m, Røste (?): c. 1 m.

Eoplacognathus suecicus Zone (Bergström 1971)

Original definition. – Originally, Bergström (1971, p. 94) defined the *E. suecicus* Zone as the *Pygodus serra – E. suecicus* Subzone. The base was marked by the appearance of *P. serra* and the top by the appearance of *Eoplacognathus foliaceus*. However, in subsequent papers (e.g. Löfgren 1978; Bergström 1983; Zhang & Sturkell 1998) the interval has been referred to as the *E. suecicus* Zone, and the lower boundary of the *P. serra* Zone was moved to near the base of succeeding the *E. foliaceus* Subzone.

Emended definition. – Interval between the first appearance of *E. suecicus* and the first appearance of *Pygodus serra*.

Characteristics. – At Andersön-B, the basal part of the subzone (sample 99478) is dominated by *Protopanderodus robustus*, but also *Periodon macrodentata*, *Minimodus poulseni* and *Protopanderodus graeai* are common. *E. suecicus* is present to rare. The lowermost sample of the *E. suecicus* Zone at Steinsodden contained only 19 specimens (sample 69673), but also here *P. robustus* is the most frequent. The high abundance of *P. robustus* is even more evident at Røste. *Costiconus ethingtoni* appears close to the base of the *E. suecicus* Zone at both Steinsodden, Andersön and Røste, and its appearance is an useful indicator for the base of the *E. suecicus* Zone. The overlying sample at Andersön-B is also dominated by *P. robustus*, but in contrast to the underlying interval, *Costiconus ethingtoni* and *Histiodella kristinae* are fairly common. The *Histiodella holodentata* – *H. kristinae* turnover, however, takes place just below the base of the *E. suecicus* Zone (Steinsodden, sample 69667).

The top of the Stein Formation both at Andersön-B and Steinsodden shows a considerable increase in the relative number of *Periodon macrodentata* caused by the transgression, but also this level is characterised by fairly high numbers of *P. robustus*, *P. graeai* and *C. ethingtoni*. The limestone nodules from the overlying Elnes Formation at Steinsodden are greatly dominated by *Periodon macrodentata* and *P. aculeatus*.

Distribution and thickness. – Andersön-B: +2.0 m; Steinsodden: +1.4 m; Røste: c. 1 m.

Remarks. – Two subzones of the *E. suecicus* Zone can be recognised in the autochthonous successions of Jämtland (Löfgren 1978), where the lower *E. suecicus* – *Scalpellodus gracilis* Subzone is succeeded by the *E. suecicus* – *Panderodus sulcatus* Subzone. It was not possible to distinguish the proposed subzones of the *E. suecicus* Zone in the Andersön-B section, since the three species co-occur throughout the zone. Zhang (1998a) subdivided the *E. suecicus* Zone in the *Pygodus lunnensis* and *P. anitae* subzones based on sections from central Sweden (Lunne and Hällekis). None of the two subzonal index-species have been observed in the present study. Consequently, the *E. suecicus* Zone have not been divided into subzones in the present paper.

Pygodus serra Interval Zone (Bergström 1971, emended herein)

Original definition. – The *Pygodus serra* Zone was defined as "...the stratigraphic range of the multielement species *Pygodus serrus* (Hadding). Its top is marked by the appearance of *Pygodus anserinus*..." (Bergström 1971, p. 91).

Emended definition. – Interval from the first appearance of *Pygodus serra* (Hadding) to the first appearance of *Pygodus anserinus* Lamont & Lindström.

Distribution and thickness. – Andersön-B, unknown thickness (more than 1.5 m).

Characteristics. – Sample 99484 is dominated by *Periodon aculeatus* and *Pygodus serra*, while *E. reclinatus* occurs sparsely.

Remarks. – It has been shown that *P. serra* and *P. anserinus* have overlapping ranges within the lower part of the *P. anserinus* Zone (Fåhræus & Hunter 1981). Accordingly, the zonal definition is modified here. Bergström (1983, modified after Bergström 1971) divided the *P. serra* Zone into four subzones. Of these, only the *P. serra* – *Eoplacognathus reclinatus* Subzone has been recognised in the present study. However, this may be due to the relatively wide sampling intervals within the "Andersö Shale" succeeding the Stein Formation at Andersön.

Pygodus serra – *Eoplacognathus reclinatus* Subzone (Bergström 1971)

Original definition. – The subzone was described as the interval between the appearance of *Eoplacognathus reclinatus* (Fåhræus) and the apperance of *Eoplacognathus robustus* (Bergström) (Bergström 1971).

Characteristics. – *E. reclinatus* is relatively common in sample 99484, which is greatly dominated by *Periodon aculeatus* and *Pygodus serra*. One *Amorphognathus*? Pa element was observed.

Distribution and thickness. – Andersön-B, unknown thickness (more than 0.1 m).

Pygodus anserinus Zone (Bergström 1971)

Original definition. – The base of the subzone coincides with "..the level at which *Pygodus serrus* evolves into *P. anserinus*", while the top is marked by the appearance of *Amorphognathus tvaerensis* (Bergström 1971, p. 97).

Characteristics. – The fauna from Andersön is characterised by *Pygodus anserinus*, "*Drepanoistodus venustus*", *Protopanderodus robustus*, *Dapsilodus? viruensis* and *Costiconus* sp. A and *Periodon aculeatus*. *Scabbardella altipes* dominates the conodont fauna at Høyberget, but also *Pygodus anserinus* and *Baltoniodus variabilis* are common.

Distribution and thickness. – Andersön-B: +2.1 m; Høyberget: unknown thickness; Engerdal: unknown thickness; Glöte: unknown thickness; Sorken(?): unknown thickness.

Remarks. – The *Pygodus anserinus* Zone has been divided into the *P. anserinus-Sagittodontina kielcensis* (= *Amorphognathus kielcensis* of Bergström 1983) Subzone and the overlying *P. anserinus-Amorphognathus inaequalis* Subzone (Bergström 1983). According to Bergström (1983, p. 50) the subzonal transition occurs at approximately the same stratigraphical level as the transition from *Baltoniodus prevariabilis* to *Baltoniodus variabilis*. Conodonts from the *Pygodus anserinus* Zone at Andersön have earlier been reported by Bergström (1971) and Bergström *et al.* (1974).

Pygodus anserinus – Sagittodontina kielcensis Subzone (Bergström 1983)

Original definition. – Interval between the appearance of *Pygodus anserinus* and the appearance of *Amorphognathus inaequalis* (Bergström 1983).

Characteristics. – The first appearance of *Pygodus anserinus* is within the uppermost part of the lower "Andersö shale" at Andersön (sample 97485). The fauna is characterised by the index species in association with "*Drepanoistodus venustus*" and *Protopanderodus robustus*. The next higher sample of this subzone is from the base of the "Biseriata Limestone". It is dominated by *Dapsilodus? viruensis* and *Costiconus* sp. A, while *Pygodus anserinus*, *Periodon aculeatus* and *Protopanderodus robustus* are fairly common. *Eoplacognathus lindstroemi* occurs sporadically.

Sample 99488 from the top of the "Biseriata Limestone" at Andersön-B is dominated by *Pygodus anserinus*, whereas "*Drepanoistodus venustus*" and *Protopanderodus graeai* are relatively common. The presence of advanced specimens of *Baltoniodus prevariabilis* indicate that this sample possibly correlates with a level close to the succeeding *Pygodus anserinus-Amorphognathus inaequalis* Subzone.

Distribution and thickness. – Andersön-B: +1.5 m.

Pygodus anserinus – Amorphognathus inaequalis Subzone (Bergström 1983)

Definition. – Interval between the appearance of *Amorphognathus inaequalis* and the appearance of *Amorphognathus tvaerensis* (Bergström 1983).

Characteristics. – *Scabbardella altipes* dominates the fauna at Høyberget (Rasmussen & Stouge 1989) and

Engerdalen, but also *Pygodus anserinus* and *Periodon aculeatus* are common. *Eoplacognathus* sp. indet. and *Baltoniodus variabilis* occur in low numbers.

A reference sample from the autochthonous succession at Glöte, Härjedalen, Sweden contained fairly many "*Drepanoistodus venustus*" and *Protopanderodus robustus* associated with *Pygodus anserinus* and *Baltoniodus variabilis*.

Distribution and thickness. – Høyberget: unknown thickness; Engerdal: unknown thickness; Glöte: unknown thickness; Sorken?: unknown thickness.

Remarks. – The lower boundary is typically simultaneous with the first occurrence of *Baltoniodus variabilis* (Bergström 1983).

Correlation and discussion of conodont zones

Introduction

The investigated platform margin succession has been correlated with several different key localities or areas, which in most cases faces the Iapetus Ocean. The correlation with the Baltoscandian platform zonation (Lindström 1971; Bergström 1971; van Wamel 1974; Löfgren 1978, 1993a, 1993b, 1994, 1995b; Stouge 1989; Bagnoli & Stouge 1997 and Zhang 1998a) has been carried out in particular detail. See Löfgren (1978) and Stouge (1984) for additional international correlations.

Baltoscandia

The first attempts to establish an Ordovician conodont zonation of the Baltoscandic area were by Lindström (1955, 1960); Sergeeva (1964) and Viira (1966), which were based on a form-taxonomic framework. Lindström (1971) introduced several new zones for the Lower Ordovician, while Bergström (1971) established the conodont zonation for the Middle and Upper Ordovician (see Fig. 28). Both zonations were based on the multielemental taxonomic concept. The *A. variabilis* Zone of Lindström (1971) and the succeeding *E. suecicus* Zone of Bergström (1971) were subdivided into a total of four subzones by Löfgren (1978). Van Wamel (1974) proposed a detailed conodont zonation of the Tremadoc and Arenig carbonate succession of Öland, south Sweden. The Lower Ordovician conodont zonation was reviewed by Stouge (1989), and a revised conodont zonation covering the Lower Ordovician succession of Öland, was proposed by Stouge & Bagnoli (1990) and Bagnoli & Stouge (1997). Löfgren (1993a, b, 1994, 1995) subdivided both the *P. proteus*, *O. evae*, *B.*

BALTIC STAGES AND SUBSTAGES Jaanusson (1982); Männil (1990); Lindholm (1991)	BALTOSCANDIAN PLATFORM MARGIN This paper — Zones	Subzones	BALTOSCANDIAN PLATFORM Lindström (1971); Bergström (1971, 1986); Löfgren (1978) — Zones	Subzones	CENTRAL SWEDEN Löfgren (1993a, b, 1994, 1995) — Zones	Subunits	ÖLAND, SWEDEN Bagnoli & Stouge (1997)	WESTERN NEWFOUNDLAND Stouge & Bagnoli (1988, 1997); E. Ord.; Stouge (1984); M. Ordovician	NORTH AMERICAN MIDCONTINENT compiled by Webby (1995), mainly after Ethington & Clark (1982) and Sweet (1984)
KUKRUSEAN	P. anserinus	A. inaequalis / A. kielcensis	P. anserinus	A. inaequalis / A. kielcensis					C. sweeti
UHAKUAN	P. serra	E. lindstroemi (not observed) / E. robustus (not observed) / E. reclinatus	P. serra	E. lindstroemi / E. robustus / E. reclinatus					C. friendsvillensis
LASNAMÄGIAN		E. follaceus (not observed)		E. follaceus				Not zoned	P. "pre-flexuosus"
ASERIAN	E. suecicus	E. suecicus	E. suecicus	E. suecicus - P. sulcatus / E. suecicus - S. gracilis				H. kristinae	
KUNDAN	P. graeai / M. ozarkodella / B. medius - H. holodentata	E? variabilis / M. ozarkodella / E? variabilis / M. flab. parva	E? variabilis	E? variabilis / M. ozarkodella / E? variabilis / M. flab. parva	Not zoned		Lenodus sp. A / L. antivariabilis / B. norrlandicus	H. tableheadensis / Not zoned	H. holodentata / H. sinuosa
	B. norrlandicus - D. stougei	M. flab. parva	M. flab. parva		M. parva		M. parva	Not zoned	H. altifrons
VOLKHOVIAN	P. rectus - M. parva / P. zgierzensis		P. originalis		P. originalis	PHASE 5 / PHASE 4 / PHASE 3 / PHASE 2 / PHASE 1	B. navis	Not zoned	M. flabellum - T. laevis
	M. flabellum - D. forceps		B. navis / B. triangularis		B. navis, lower, middle, upper / B. triangularis	upper - middle int. / lower - middle int. / lower interval	M. flabellum / B. ? triangularis / Microzark. sp. A / T. diprion / T. triangularis		
BILLINGENIAN	O. evae		O. evae		O. evae		O. evae	O. evae	R. andinus / O. communis
	P. elegans		P. elegans		P. elegans		P. elegans	P. elegans	A. deltatus - O. costatus
HUNNEBERGIAN	P. proteus		P. proteus		P. proteus	O. elongatus - A. d. deltatus / P. gracilis / Tripodus / D. aff. amoenus	Not zoned	P. adami / P. oepiki / Barren interzone / P. gliberti	M. dianae / "Low diversity"
TREMADOC (pars)	P. deltifer		P. deltifer		P. deltifer			Not zoned	R. manitouensis

Fig. 28. Correlation of the Baltoscandian platform margin conodont zones with the conodont zonations of the Baltoscandian platform and Newfoundland, and the North American standard zonation.

triangularis, *B. navis* and *P. originalis* zones of the central Sweden.

Correlation with the zonation of Lindström (1971) and Bergström (1971) and the subsequent modifications by Löfgren (1978) and Zhang (1998a)

Paltodus deltifer Zone (Lindström 1971). – The Grøslii samples contain *Paltodus* cf. *deltifer* (Lindström), *Paroistodus numarcuatus* (Lindström) and "*Scolopodus*" *peselephantis* Lindström. Both these samples and the sample 97757 from Røste, which contain *P. deltifer* and *Paroistodus numarcuatus* correlate with the *P. deltifer* Zone of Lindström (1971). A relatively rich conodont fauna from this zone at Grøslii was reported by Repetski (in Bruton *et al.* 1989). The fauna is similar to the faunas described from the corresponding zone on the proximal platform (Lindström 1955, 1971; Löfgren 1978; Olgun 1987 and Bagnoli *et al.* 1988), although the occurrences of *Cordylodus*, which was reported by Lindström (1971) and Sturkell (1991) from the lower part of the "*Ceratopyge Limestone*" was not observed in the few samples from the *P. deltifer* Zone studied in the present work.

Paroistodus proteus Zone (Lindström 1971). – The basal sample of the Andersön-B section (99451) is dominated by *Paracordylodus gracilis* Lindström, *Paroistodus proteus* and *Tetraprioniodus robustus* Lindström. A few specimens of *Paltodus subaequalis* and *Scolopodus quadratus* Pander 1856 were also recorded. The appearance of *Oelandodus elongatus* van Wamel 1974, *Paracordylodus gracilis*, *Paroistodus proteus* and/or *P. parallelus* (Pander 1856) in the overlying sample 99452, indicate that the Tremadoc-Arenig boundary is placed just below this sample. The same three species were observed in three lowermost samples from the Andersön-A section (97800, 97801, 97802), which also correlate with the *P. proteus* Zone as defined by Lindström (1971) (Fig. 28). The samples from the Bjørkåsholmen Formation at Steinsodden yielded *Paroistodus numarcuatus*, *Paltodus* cf. *subaequalis* (Pander), *Paroistodus proteus* (Lindström) and *Paltodus deltifer* (only the lower sample), and correlate with the lower part of the *P. proteus* Zone of Lindström (1971).

Prioniodus elegans Zone (Lindström 1971). – *Prioniodus elegans* Pander 1856 dominates the sparse conodont fauna within a thin limestone bed of the Tøyen Formation at Andersön-A (sample 99803) and thus correlates with the *P. elegans* Zone *sensu* Lindström (1971). Sample 99453 from the Andersön-B section lacks *Prioniodus elegans* but the occurrence of *Oelandodus elongatus*, *Paracordylodus gracilis*, *Paroistodus parallelus* and *Stolodus stola* (Lindström) indicates correlation with the *P. elegans* Zone.

Oepikodus evae Zone (Lindström 1971). – Sample 97805 from the Andersön-A section contains *Oepikodus evae* (Lindström), *Stolodus stola*, *Paroistodus parallelus* and *Oistodus lanceolatus* Pander 1856 and correlates with the lower part of the *O. evae* Zone. Sample 99454 from the Andersön-B section comprises *Drepanoistodus forceps*, *Periodon selenopsis*, *Paroistodus originalis*, *Drepanodus arcuatus*, *Protopanderodus rectus* and *Protopanderodus calceatus*. This sample correlates with the upper part of the *O. evae* Zone, which is characterised by the lack of the nominate species on the Baltoscandian platform (Lindström 1971).

Baltoniodus triangularis and *Baltoniodus navis* zones (Lindström 1971). – The *M. flabellum – D. forceps* Zone of this study correlates approximately with the *Baltoniodus triangularis* and *Baltoniodus navis* Zones of Lindström (1971). Originally, the *Baltoniodus triangularis* Zone was defined as the interval from the appearance of *B. triangularis* (Lindström) to the appearance of *B. navis* (Lindström).

Lindström (1971) noted that *Baltoniodus triangularis* characterises the lowermost part of Volkhov, and first appears at a level close to "Blommiga-bladet" on Öland. This is a characteristic hardground with numerous prominent burrows within the Bruddesta Formation of van Wamel (1974). *Baltoniodus navis* appears a little higher in the succession while *Microzarkodina flabellum*, according to Lindström, occurs for the first time just below "Blommiga-bladet". The base of *Baltoniodus navis* Zone was placed where *Baltoniodus navis* appears for the first time. Unfortunately, *Baltoniodus* is infrequent and poorly preserved in the interval around the base of the Volkhov within the sections studied herein. This makes a precise correlation with the *B. triangularis* and *B. navis* zones difficult. In conclusion, the base of the *M. flabellum – D. forceps* Zone correlates with a level between the base of the *B. triangularis* Zone (Stouge 1989) and the base of the "upper interval" of the *B. triangularis* Zone (Löfgren 1993b).

Paroistodus originalis Zone (Lindström 1971). – The base of the *P. originalis* Zone was defined by the coincidence of three events: 1) The appearance of *Triangulodus brevibasis* (Sergeeva 1963) (= *Scandodus brevibasis* of Lindström 1971), 2) the level where *Paroistodus originalis* becomes frequent, and 3) the level where *Drepanoistodus basiovalis* (Sergeeva 1963) outnumbers *D. forceps* (Lindström 1971, p. 31–32, fig. 2). Subsequent papers by Kohut (1972); Stouge (1975, 1989), Löfgren (1978, 1985, 1995); Olgun (1987); Stouge (1989); Stouge & Bagnoli (1990) and Rasmussen (1991) have contributed new data and several modifications and new interpretations of the *P. originalis* Zone. Because the vertical and lateral ranges and faunal dynamics of the species, which were originally used to define the *P. originalis* Zone by Lindström (1971), are clearly different within the sections studied here, a new zone, the *Periodon zgierzensis* Zone, is established for the lower part of the interval. The following major differences occur between the *P. originalis* Zone as originally defined and the corresponding interval studied here: 1) *P. originalis* becomes abundant already in the interval corresponding to the *B. navis* Zone. This is particularly distinct within the Herram section, but it is also known from the Komstad Formation, Bornholm, Denmark (Stouge 1975); 2) *P. originalis* ranges into the Llanvirn, and 3) *Triangulodus brevibasis* has a diachronous appearance across the Baltoscandian platform and also between the Andersön sections.

The base of the *P. rectus–M. parva* Zone of the present paper is taken at the level where *M. parva* appears for the first time. This correlates with the level 1.5 m above the base of the *P. originalis* Zone at Finngrundet, south Bothnian Bay (Löfgren 1985); just below the base of the *P. originalis* Zone at northern Öland (Stouge & Bagnoli 1990) and at the base of the *M. parva* Zone (*sensu* Stouge 1989) at Slemmestad, Oslo (Rasmussen 1991). The upper boundary of the *P. rectus–M. parva* Zone as defined here correlates roughly with the base of the "classical" *Microzarkodina parva* Zone as defined by the same authors.

Microzarkodina parva Zone (Lindström 1971). – Lindström (1971) defined the *Microzarkodina parva* Zone as the interval where "*Semiacontiodus*" *cornuformis* (Sergeeva 1963) (= *Protopanderodus cornuformis* sensu Lindström 1971) coexists with *Microzarkodina parva* Lindström 1971 and *Triangulodus brevibasis* (Lindström 1971, p. 32). Because these species have somewhat different vertical ranges at different localities, the definition has later been modified and changed several times: Interval between the appearance of "*Semiacontiodus*" *cornuformis* and *Eoplacognathus? variabilis* (Sergeeva 1963) (Löfgren 1978, 1985); interval between the appearance of *Baltoniodus norrlandicus* Löfgren 1978 and *Eoplacognathus? variabilis* (Stouge 1989; Rasmussen 1991); interval between the appearance of *Baltoniodus norrlandicus*, *Drepanoistodus basiovalis*, *Protopanderodus* cf. *varicostatus* (Sweet & Bergström 1962) and "*Semiacontiodus*" *cornuformis* and the appearance of *Lenodus* n. sp. A (Stouge & Bagnoli 1990).

The base of the *B. norrlandicus – D. stougei* Zone coincides approximately with the base of the *Microzarkodina parva* Zone as interpreted by Löfgren (1978, 1985); Stouge (1989); Stouge & Bagnoli (1990) and Rasmussen (1991) (see above). According to these authors *Baltoniodus norrlandicus* and "*Semiacontiodus*" *cornuformis* appear at about the same level in the Central Baltoscandian Confacies Belt while the latter species appears clearly later closer to the platform margin.

Amorphognathus variabilis Zone (Lindström 1971, modified by Löfgren 1978). – Lindström (1971, p. 32) reported that the Hunderum Substage is distinguished by the presence of *Amorphognathus variabilis* in most samples, and correlated the *A. variabilis* Zone with the Kundan Stage (Lindström 1971, fig. 1). Subsequently, Löfgren (1978) subdivided the *Eoplacognathus? variabilis* Zone (= *A. variabilis* zone of Lindström 1971) into the *E.? variabilis - M. flabllum* (lower) and *E.? variabilis – M. ozarkodella* (upper) subzones. The upper boundary of the *B. norrlandicus – D. stougei* Zone of the present paper correlates with a level somewhere in the lower part of the *Amorphognathus variabilis* Zone of Lindström (1971) and *E.? variabilis – M. flabellum* Subzone of Löfgren (1978). However, as the platform elements referred to *Eoplacognathus? variabilis* by for example Löfgren (1978) and Rasmussen (1991) are in need of revision, it has not been possible to make a precise correlation with the base of the *Amorphognathus variabilis* Zone. The studies by Stouge & Bagnoli (1990) and Zhang (1997) have improved the knowledge of the Baltoscandian, late Arenig and early Llanvirn platform conodont species considerably, but because of the low number of *Lenodus* and *Eoplacognathus* specimens described herein, it has not been possible to use the new interpretations.

The "*Baltoniodus medius* zone" was used informally by Stouge (1989). Based primarily on studies of Swedish shelf sections, he reported that *Baltoniodus medius* appears at the same level as "*Drepanoistodus venustus* (Stauffer 1935)", *Polonodus* spp. and *Ansella jemtlandica* (Löfgren 1978). The base of the *B. medius– H. holodentata* Zone (this paper), which indicates the beginning of a transgressive phase, correlates approximately with this level. Accordingly, the *B. medius–H. holodentata* Zone correlates with the interval within the middle part of the *Amorphognathus variabilis* Zone of Lindström (1971). The base of the *B. medius–H. holodentata* Zone approximately coincides with the Arenig–Llanvirn boundary (see discussion by Rasmussen 1991).

Microzarkodina ozarkodella Lindström 1971 outnumbers *Microzarkodina hagetiana* Stouge & Bagnoli 1990 at the base of the *L. variabilis–M. ozarkodella* Subzone of Löfgren (1978), which correponds to base of the *M. ozarkodella* Zone on the present paper. *Protopanderodus graeai* appears in the upper part of the *L. variabilis– M. ozarkodella* Subzone in the autochthon of Jämtland (Löfgren 1978). This makes the interval corresponding to the *P. graeai* Zone relatively thin compared to the development within the platform margin sections studied here.

Eoplacognathus suecicus Zone (Bergström 1971).– Bergström (1971) placed the base of the zone where *Pygodus serra* appears for the first time and the top where *Eoplacognathus foliaceus* appears. However, in subsequent papers (e.g. Löfgren 1978; Bergström 1983; Zhang & Sturkell 1998) the interval has been referred to as the *E. suecicus* Zone, and the lower boundary of the *P. serra* Zone was moved to near the base of succeeding the *E. foliaceus* Subzone. It was not possible to use the *E. suecicus– Scalpellodus gracilis* and *E. suecicus–P. sulcatus* subzones of Löfgren (1978) or the *E. suecicus– Pygodus anitae* and *E. suecicus–Pygodus lunnensis* subzones of Zhang (1998a) in the present study. See the taxonomical description of *E. suecicus* for further details.

The *Pygodus serra* and *P. anserinus* zones (Bergström 1971).– The faunal characteristics of the *Pygodus serra* and *P. anserinus* zones and subzones within the Andersön-B section is in accordance with the description of Bergström (1971, 1983).

The faunal associations within the Elvdal Formation (Rasmussen & Stouge 1989) is slightly aberrant because of the relatively high abundance of *Scabbardella altipes* (Henningsmoen 1948) and *Periodon aculeatus* Hadding 1913 and the very few *Eoplacognathus* specimens. The presence of *Pygodus anserinus* Lamont & Lindström 1957 and *Baltoniodus variabilis* (Bergström 1962), however, indicates correlation with the *P. anserinus– A. inaequalis* Subzone.

Correlation with the conodont zone subdivisions by Löfgren (1993a, b, 1994, 1995b)

Paroistodus proteus Zone (Lindström 1971) *sensu* Löfgren (1993a, 1994).– The *Paroistodus proteus* Zone of Lindström (1971) of the autochthonous areas in Sweden was subdivided into four intervals by Löfgren (1993a, 1994) (Fig. 28). The *Drepanoistodus* aff. *D. amoenus* Subzone was characterised by the presence of *P. proteus* (more numerous than *P. numarcuatus*), *Paltodus deltifer*, *Paltodus* cf. *subaequalis* and *Acodus* aff. *deltatus*. The following interval, the *Tripodus* Subzone was distinguished by the appearance of *Paltodus subaequalis*, *Tropodus comptus australis* and *Drepanoi-*

stodus forceps, where the first occurrence of the last species defines the base of the zone. The base of the succeeding *Paracordylodus gracilis* was identified by the first occurrence of *Paracordylodus gracilis*, while the *Oelandodus elongatus– Acodus deltatus deltatus* Subzone (Löfgren 1994), that is identical with the "uppermost *P. proteus* Zone" of Löfgren (1993a) was recognised by the incoming of *Oelandodus elongatus*. It is difficult to correlate the subdivisions defined by Löfgren with the conodont succession within the present study. A distinct difference between the Swedish sections studied by Löfgren (1993a, 1994) and the Andersön sections is that *Paracordylodus gracilis* dominates the lower part of the *Oelandodus elongatus* bearing interval of the *P. proteus* Zone at Andersön. It is much more common here than in the contemporary shelf areas studied by Löfgren.

The lowermost sample from the Bjørkåsholmen Formation at Steinsodden contains among others *Paroistodus numarcuatus*, *Paltodus* cf. *subaequalis*, *Paroistodus proteus* and *Paltodus deltifer*. The sample is part of the *P. proteus* Zone as defined here, but correlates with the uppermost part of the *P. deltifer* Zone *sensu* Löfgren (1993a). She placed the base of the *P. proteus* Zone where the nominate species out-numbers *P. numarcuatus*. Subsequently, Löfgren (1994) redefined the base of the *P. proteus* Zone to the level where *Paroistodus proteus* appears for the first time.

The basal sample at the Andersön-B section (99451) is dominated by *Paracordylodus gracilis*, *Paroistodus proteus* and *Tetraprioniodus robustus* together with low frequencies of *Paltodus subaequalis* and *Scolopodus quadratus*. This assemblage indicates correlation with the *Paracordylodus gracilis* Subzone of the *P. proteus* Zone of Löfgren (1994). The presence of *Oelandodus elongatus*, *Paracordylodus gracilis*, *Paroistodus proteus* and/or *Paroistodus parallelus* in the overlying sample 99452 and the three lowermost samples from the Andersön-A section (97800, 97801, 97802) indicates correlation with the *O. elongatus-A. deltatus deltatus* Subzone (Löfgren 1994) of the *P. proteus* Zone.

Prioniodus elegans Zone (Lindström 1971) *sensu* Löfgren (1993a, b, 1994).– The faunal composition of sample 97804 from Andersön-A, for example *Prioniodus elegans*, is distinctive for the lower part of the *P. elegans* Zone of Löfgren (1993a, b, 1994).

Oepikodus evae Zone (Lindström 1971) *sensu* Löfgren (1993b).– The *O. evae* Zone of the autochthonous shelf areas in Sweden was divided into four intervals by Löfgren (1993b). The lower interval was characterised by relatively low frequencies of *D. forceps* and *Protopanderodus rectus* and the lack of *Oistodus lanceolatus*. The latter species appeared in the lower–

middle interval which was also characterised by a conspicuous increase in the frequency of *P. rectus*. The occurrence of *Nordiora? crassulus* and *O. evae* typified the upper–middle interval, while the upper interval was characterised by the lack of *Paroistodus parallelus* and *O. evae*.

Sample 97805 from Andersön-A contains *Drepanoistodus forceps*, *Oepikodus evae*, *Paroistodus parallelus* and *Oistodus lanceolatus* and probably correlates with the lower–middle or upper–middle interval of Löfgren (1993b). Sample 99454 from the Andersön-B section was characterised by the occurrence of *Drepanoistodus forceps*, *Periodon selenopsis*, *Paroistodus originalis*, *Drepanodus arcuatus*, and *Protopanderodus rectus*, whilst *O. evae* is lacking. The fauna is typical for the upper part of the *O. evae* Zone *sensu* Löfgren (1993b).

Baltoniodus triangularis and *Baltoniodus navis* zones (Lindström 1971) *sensu* Löfgren (1993b). – The *B. triangularis* Zone was divided into two intervals by Löfgren (1993b). The lower *B. triangularis* interval was defined by the first occurrence of *B. triangularis* (Lindström 1955), and the upper interval by the appearance of *Microzarkodina flabellum* (Lindström 1955).

Löfgren (1993b) subdivided the overlying *B. navis* Zone into three intervals: The lower interval was characterised by the appearance of *B. navis*, the middle by the appearance of *Lenodus? sp. A* (which is different from *Lenodus sp. A* in this paper and probably belongs to *Trapezognathus quadrangulum*) and the upper interval by the appearance of *Triangulodus brevibasis*. It is not possible to recognise the three *B. navis* Zone intervals defined by Löfgren (1993b) in the allochthonous sequences as the nominate taxa are rarely represented and the appearance of *T. brevibasis* is diachronous. *T. brevibasis* appears in the middle part of the *M. flabellum – D. forceps* Zone at Andersön-B, whereas it appears within the *P. rectus – M. parva* Zone in the Andersön-A section. *T. brevibasis* has not been recorded from the Heramb and Steinsodden sections.

Paroistodus originalis Zone (Lindström 1971) *sensu* Löfgren (1995b). – The *Paroistodus originalis* Zone was subdivided into five phases by Löfgren (1995b) based on samples from various localities within the central Baltoscandian confacies belt. Each of the five phases was characterised by different conodont assemblages. The spatial distribution and relative number of many of genera and species that define the individual phases are now known to be related to changes in the palaeoenvironments (Rasmussen & Stouge 1995). It means that the distribution, and especially the acmes, of several of the taxa in question varied considerably across the Baltoscandian platform. Because of this, it is difficult to correlate the five phases of the *P. originalis* Zone from the relatively shallow-water deposits of the central Baltoscandian confacies belt with the platform margin zones established herein. However, Löfgren (1995b) showed that the phases are recognisable in the main part of the Central Baltoscandian Confacies Belt, and therefore may help to refine the lateral correlations within this area.

Correlation with the conodont zonation by van Wamel (1974), Öland, Sweden

Van Wamel (1974) introduced a very detailed conodont zonation for the Tremadoc, Hunnebergian, Billingenian and lower Volkhovian succession of Öland, southeast Sweden. Although van Wamel's taxonomic interpretations to some extent differ from the one presented here, it is possible to correlate most of the platform margin zones with the Öland succession. However, several of van Wamel's zones have not been distinguished in the present study, partly because the species have different stratigraphical ranges, and partly because the interval studied by van Wamel (1974) is dominated by shales on the platform margin and thus contains relatively few conodont specimens.

The *P. deltifer* Zone strata recognised at the Grøslii and Røste sections comprise van Wamel's *Drepanoistodus numarcuatus – Paroistodus amoenus* and *Drepanodus arcuatus* zones. The *Prioniodus deltatus* Zone correlates with the lower part of the *P. proteus* Interval Zone of the present paper. *Drepanoistodus inconstans* (Lindström 1955) *sensu* van Wamel 1974 is partly synonymous with *Paltodus subaequalis* Pander 1856, and *G. microdentatus* van Wamel 1974 of *Tetraprioniodus robustus* Lindström 1955. As a consequence, the *D. inconstans* and *T. subtilis – G. microdentatus* zones of van Wamel correspond to the *P. proteus* Zone at Andersön-A and -B.

The *Prioniodus elegans* Zone as defined here is the interval from the first occurrence of *P. elegans* to the first appearance of *Oepikodus evae*. Accordingly, it correlates with the van Wamel's *P. elegans – Oelandodus elongatus*, *Stolodus stola* and *Protopanderodus rectus – Oelandodus costatus* zones. The *O. evae* Zone as defined in this work – that is the interval between the first occurrence of *O. evae* and the first occurrence of *Microzarkodina flabellum* – apparently correlates with seven van Wamel zones: The *Protopanderodus rectus – Oepikodus evae*, *Oistodus lanceolatus – Prioniodus deltatus*, *Oistodus lanceolatus*, *Prioniodus crassulus*, *Prioniodus navis – Prioniodus crassulus*, *Prioniodus navis – Stolodus stola* and *Drepanoistodus forceps* zones, although the individual zones cannot be distinguished within the platform margin successions studied here.

The first occurrence of *Prioniodus navis sensu* van Wamel (1974) (= *Trapezognathus diprion sensu* Bagnoli & Stouge 1997) is at a level within the Köpingsklint Formation at Köpingsklint, and within the lower part of the Bruddesta Formation at Äleklinta – Djupvik and Horns Udde (van Wamel 1974, chart 2, 3, 4). At all localities, "*P.*" *navis* co-occurs with "*Prioniodus*" *crassulus* at a level significantly below "Blommigabladet". Accordingly, it may be concluded that the interval from the *Protopanderodus rectus – Oepikodus evae* Zone to the top of *Drepanoistodus forceps* Zone in the zonation scheme by van Wamel (1974) correlates largely with the *Oepikodus evae* Interval Zone as defined herein. Van Wamel's *Microzarkodina flabellum* Zone was defined as the interval between the appearance of the nominate species and the appearance of *Triangulodus brevibasis*. Bagnoli & Stouge (1997) showed that an early *Microzarkodina* species (*Microzarkodina* sp. A *sensu* Bagnoli & Stouge) appears before the *Baltoniodus triangularis* Zone. Therefore, the basal part of van Wamel's *Microzarkodina flabellum* Zone probably correlates with the uppermost part of the *Oepikodus evae* Zone *sensu* this paper. The succeeding part of the *M. flabellum* Zone and the lower part of his overlying *Triangulodus brevibasis* Zone (the upper boundary was not defined) correlates with the *M. flabellum – D. forceps* Zone of the present study.

Correlation with the conodont zonation by Bagnoli & Stouge (1997), northern Öland, Sweden

Bagnoli & Stouge (1997) studied the late Latorpian to early Kundan conodont succession at Horns Udde, northern Öland, and proposed a new conodont zonation for this part of the Lower Ordovician. At this time, northern Öland was part of the central Baltoscandian confacies belt or even the north Estonian confacies belt, which means that the stratigraphical range of several conodont species and genera are different from their range in the distal parts of the platform farther west. Some of the zonal boundaries, however, are similar to those which have been defined herein, meaning that it is possible to correlate closely between northern Öland and the Scandinavian mountain belt at certain stratigraphic levels.

The *Oepikodus evae* interval Zone of Bagnoli & Stouge (1997) correlates with the lower and middle part of the *O. evae* Zone as defined herein. Sample 97805 from the Andersön-A section correlates with a level within this interval zone.

The following *Trapezognathus diprion* interval Zone

was defined as the interval between the first occurrence of the nominate species and the first occurrence of *Microzarkodina* sp. A Bagnoli & Stouge 1997. It is possible that sample 99454 from the upper part of the *O. evae* Zone (*sensu* this paper) at the Andersön-B section, which comprises *Drepanoistodus forceps*, *Periodon selenopsis*, *Paroistodus originalis*, *Drepanodus arcuatus*, *Protopanderodus rectus* and the *P. sulcatus/ P. calceatus* complex correlates with this zone.

The appearance of *Microzarkodina* n. sp. A Bagnoli & Stouge 1997 marks the base of the *Microzarkodina* n. sp. A interval Zone. *Microzarkodina* n. sp. A has not been observed in the present study, and the correlation with the distal platform sections seems difficult at the present. Bagnoli & Stouge (1997) correlated the zone with the uppermost part of the *O. evae* Zone *sensu* Lindström (1971).

The following *Baltoniodus*? *triangularis* interval Zone was defined by Bagnoli & Stouge (1997) as the interval between the first appearance of *B.*? *triangularis* and the first apperance of *Microzarkodina flabellum*. Because of the generally poor preservation of *Baltoniodus* elements in the material at hand, it is almost impossible to distinguish the elements of the *B. triangularis – Trapezognathus quadrangulum* complex from *B. navis* in the sections studied herein, meaning that the identification of the zone is difficult.

The *Microzarkodina flabellum* interval Zone was defined as the interval between the first appearance of the nominate species and the appearance of *Baltoniodus navis*. Thus, the base of the zone is identical to the base of the *Microzarkodina flabellum – Drepanoistodus forceps* Concurrent Range Zone in the present paper. It is assumed that the base of the *B. navis* Zone of Bagnoli & Stouge correlates with a level within the lower part of the *M. flabellum – D. forceps* Concurrent Range Zone, based on the upper range of *D. forceps*. *Oistodus lanceolatus* has its upper range at the top of the *M. flabellum* Zone at Horns Udde, while it ranges up into the *Microzarkodina flabellum – Drepanoistodus forceps* Zone at the Andersön-B section and the *Periodon zgierzensis* Zone at the Andersön-A section. Thus, the spatial distribution of this species seems influenced by palaeoenvironmental differences.

The *Baltoniodus navis* interval Zone of Bagnoli & Stouge (1997) was defined as the interval between the first appearance of *B. navis* and the first appearance of *Microzarkodina parva*. The zone correlates with the upper part of the *M. flabellum – D. forceps* Concurrent Range Zone and the *Periodon zgierzensis* Partial Range Zone.

The overlying *Microzarkodina parva* Zone was defined by Bagnoli & Stouge (1997) as the interval between the first appearance of *M. parva* and the first appearance of *Baltoniodus norrlandicus*. It correlates with the *P. rectus–M. parva* Interval Zone as it is defined herein.

The *Baltoniodus norrlandicus* interval Zone was defined as the interval from the first appearance of *B. norrlandicus* and the first appearance of *Lenodus antivariabilis*. Based on the appearance of the nominate species and *Scalpellodus gracilis*, the zone correlates with the lower part of the *B. norrlandicus–D. stougei* Zone of the present paper.

The base of the following *Lenodus antivariabilis* interval Zone was defined by the first appearance of the nominate species, and the top as the level where *Lenodus* sp. A appears for the first time. The overlying *Lenodus* sp. A interval Zone was defined as the interval between the first appearance of the nominate species and the first appearance of *Microzarkodina hagetiana*. Neither *L. antivariabilis* nor *Lenodus* sp. A were observed in the present study but based on the ranges of e.g. *Scalpellodus gracilis* and *Microzarkodina hagetiana*, the two zones correlate with the upper part of the *B. norrlandicus–D. stougei* Zone and the lowermost part of the *B. medius–H. holodentata* zone as defined in this paper.

Poland

A number of Polish conodont studies have concentrated on erratic boulders containing typical Baltic faunas (for example Wolska 1961; Dzik 1976). However, the over 8 m thick Mójcza Limestone of the Holy Cross Mountains of southern Poland contains a diverse conodont fauna, which has several taxa in common with the Baltoscandian platform margin. Ordovician conodonts of Baltoscandian affinity have also been described from the subsurface of northern Poland (Bednarczyk 1998).

The Mójcza fauna was described by Dzik (1978, 1990, 1994). The lower approximately 4.5 m was correlated with the *A. variabilis*, *P. serra* and *P. anserinus* zones by Dzik (1990, Fig. 8). The about 1.5 m thick *A. variabilis* Zone represents the lowermost part of the section, and contains a fauna that is very similar to faunas obtained from the corresponding interval within the Stein Formation.

Nordiora sp. [= *Phragmodus polonicus* Dzik "early form", cf. Dzik 1994], *Baltoniodus medius* (Dzik) [= *Baltoniodus parvidentatus* (Sergeeva) and *B. medius sensu* Dzik 1993], and *Protopanderodus* cf. *robustus* (Hadding) [= *Protopanderodus rectus* (Lindström) cf. Dzik 1990, 1994] occur through the *A. variabilis* Zone, whilst *Microzarkodina ozarkodella* Lindström appears about 90 cm below the top (in short BT) of the zone (Dzik 1978), which is marked by a distinct discontinuity surface which also represent a prominent hiatus. *Protopanderodus graeai* (Hamar) appears about 80 cm

BT (Dzik 1990, 1994). *Histiodella kristinae* appears some 50 cm BT, indicating that this part correlates with the *E. suecicus* Zone. This suggests that the interval from the base to the level 80 cm BT correlates with the *B. medius–H. holodentata* Zone, the 90 cm–80 cm BT interval with the *M. ozarkodella* Zone, the 80 cm–50 cm BT interval with the *P. graeai* Zone and the upper 50 cm of the *A. variabilis* Zone of Dzik (1990) with the *E. suecicus* Zone of the Baltoscandian platform margin. This favours a Llanvirn age for the basal part of the section rather than the Upper Arenig age as suggested by Dzik (1994).

The faunas described from the *P. serra* and *P. anserinus* zones by Dzik (1978, 1990) share several species with the Scandinavian Lower Allochthon. The Mójcza fauna, however, comprises common *Panderodus*, *Sagittodontina* and *Complexodus*. The two first-mentioned genera are rare within the Lower Allochthon of the Scandinavian Caledonides, while *Complexodus* seems to be missing.

A rich but poorly preserved fauna was obtained from the Rzeszòwek slates in the Kaczawa Mountains, the Sudetes (Urbanek & Baranowski 1986; Dzik 1990). The fauna is not considered in detail here, but the occurrence of for example *Plectodina* and *Icriodella cerata* (Knüpfer) together with the typical Baltic taxa *Paroistodus parallelus* (Pander), *Microzarkodina flabellum* (Lindström), *Protopanderodus rectus* (Lindström) and *Cornuodus longibasis* (Lindström) reflects a "Central European" affinity of the fauna. The conodont assemblage correlates in part with the *M. flabellum–D. forceps* Zone.

Correlation with areas outside Baltoscandia

Western Newfoundland, Table Head Group

Conodonts from the Table Head Group were described by Fåhræus (1970), Stouge & Boyce (1983) and Stouge (1984) and the lithostratigraphy was revised by Klappa *et al.* (1980) and Stenzel *et al.* (1990). Stouge (1984) erected the *Histiodella tableheadensis* (regarded as *H. holodentata* in this paper) and *Histiodella kristinae* conodont zones (Fig. 28). The *H. tableheadensis* Zone characterises the main part of the Table Point Formation (at Table Point) except the uppermost part, while the *H. kristinae* Zone represents the remaining part of the Table Point Formation and all, but the uppermost part of the succeeding Table Cove Formation.

The upper boundary of the *H. kristinae* Zone is

marked by the appearance of *Histiodella bellburnensis* Stouge 1984, which characterises the uppermost part of the Table Cove Formation.

Rasmussen & Stouge (1988) demonstrated that the Stein Formation at Steinsodden contains taxa (including *Histiodella*) that enable a reliable correlation between the Baltoscandian margin and the Table Head conodont zones of the Laurentian margin. In total, the Table Head Group has fifteen species in common with the sections studied here according to the data presented by Stouge (1984).

The appearance of *Histiodella holodentata* at the base of the *H. tableheadensis* Zone of the Table Head Group indicates correlation with the base of the *B. medius – H. holodentata* Zone of the Scandinavian Lower Allochthon.

A significant faunal change occurs in sample TP36 in the middle of the *H. holodentata* Zone at the Table Point locality of Stouge (1984). *Periodon macrodentata* (*P. aculeatus zgierzensis* sensu Stouge) appears and constitutes 20% of the fauna. Also, *Ansella jemtlandica*, *Oistodus tablepointensis*, *Costiconus costatus*, *Paroistodus horridus* and *Parapaltodus simplicissimus* occur for the first time. The fauna is a typical deeper water assemblage, which correlates with the lower – middle part of the *B. medius – H. holodentata* Zone of the Baltoscandian margin, especially distinct within the Andersön-B section (Rasmussen & Stouge 1995). Between this level and a level approaching the appearance of *Histiodella kristinae*, more typical shallow water taxa dominate at both Table Point and the Baltoscandian margin.

The base of the *H. kristinae* Zone marks the beginning of a transgressive phase at both sides of the Iapetus Ocean. Within the Stein Formation, the appearance of *H. kristinae* is more or less coincident with the base of the *Eoplacognathus suecicus* Zone. The Table Head Group and the Stein Formation are dominated by *Periodon* above this level, and the present deep water assemblage was referred to the *Periodon – Cordylodus*? Biofacies by Stouge (1984). The corresponding conodont assemblage of the Stein Formation was referred to the *Protopanderodus – Periodon* Biofacies by Rasmussen & Stouge (1995).

It may be concluded that the *Histiodella holodentata* and *H. kristinae* zones of Stouge (1984) correlates with interval from the *B. medius – H. holodentata* Zone to the *E. suecicus* Zone at the Baltoscandian platform.

Western Newfoundland, Cow Head Group

The Cow Head Group comprises the Middle Cambrian – Lower Ordovician (Arenig) Shallow Bay For-

mation and the Arenig – basal Llanvirn Lower Head Formation (James & Stevens 1986). It consists predominantly of shales with intercalated limestone beds and up to several of metres thick debris flow deposits. The conodont faunas of the limestone beds were described by Fåhræus & Nowlan (1978) and Stouge & Bagnoli (1988), while Pohler *et al.* (1987) focused on the clasts within the breccias. The graptolite fauna was described by Williams & Stevens (1988).

Stouge & Bagnoli (1988) studied the interval from the uppermost part of Bed 8 to the middle part of Bed 11 (uppermost Tremadoc – lower Arenig). They erected the new *Prioniodus gilberti*, *P. oepiki* and *P. adami* zones for the interval from the uppermost part of Bed 8 to the upper-middle part of Bed 9 (Fig. 28). The *P. gilberti* Zone has *Paroistodus numarcuatus* (Stouge & Bagnoli 1988) and *Paracordylodus gracilis* (Williams *et al.* 1994; Stouge & Bagnoli 1997) in common with the sections studied herein, which suggests correlation with the *P. proteus* Zone of the Baltoscandian margin.

The base of the *P. oepiki* Zone is identical with the base of Bed 9. It is characterised by the appearance of for example *Diaphorodus tovei* Stouge & Bagnoli, *Oelandodus elongatus*, *Paroistodus proteus* and *Paroistodus parallelus*, which typifies the *P. proteus* Zone of the present study. The appearance of *Oelandodus elongatus* is contemporary with the base of the *Tetragraptus phyllograptoides* graptolite Zone (Maletz *et al.* 1996) and marks the base of the Arenig. The succeeding *P. adami* Zone has a similar fauna, except that the nominate species is the dominant species. This zone also correlates with the *P. proteus* Zone of the present study.

Prioniodus elegans is the most common species within the overlying *P. elegans* Zone at Cow Head. Both this zone and the succeeding *Oepikodus evae* Zone correlate with the corresponding zones of the Baltoscandian margin. The *O. evae* Zone at Andersön is characterised by *Periodon selenopsis* (Serpagli) and lack of *Periodon flabellum* (Lindström). Because *P. selenopsis* is much more frequent than *P. flabellum* in the lowermost part of the *O. evae* Zone at Cow Head, it is likely that the *O. evae* Zone at Andersön correlates with the lower part of the *O. evae* Zone at this locality.

Fåhræus & Nowlan (1978) reported *Periodon flabellum* and *P. aculeatus* Hadding occurring together in Bed 13 at Cow Head. The latter species includes a Pb element, which is less angulate than that of *P. aculeatus* sensu stricto, and it is more likely that it belongs to *Periodon macrodentata* (Graves & Ellison) or even its ancestor *Periodon zgierzensis*. If this is the case, Bed 13 correlates with a level within the *B. norrlandicus – D. stougei* Zone or perhaps the underlying *P. rectus – M. parva* Zone of the Baltoscandian margin.

Stouge (1984, p. 21) reported *Histiodella tableheadensis* (= *H. holodentata*) from Bed 14. This suggests that Bed 14 is likely to correlate with a level close to the base of the *B. medius–H. holodentata* Zone of the Baltoscandian margin. Stouge (in prep.) studied several samples from the overlying Bed 15, and recorded for example *H. holodentata*, *Spinodus spinatus* (Hadding), *Ansella sinuosa* (Stouge), *Paroistodus horridus* and *Oistodella* sp., which suggest correlation with *H. tableheadensis* Zone at Table Head and the *B. medius–H. holodentata* Zone herein (lowest Llanvirn).

Central Newfoundland

Prioniodus elegans was recorded from South Catchers Pond (northwestern Central Newfoundland) by Bergström *et al.* (1972) and Stouge (1980b), and *Oepikodus evae* from the Dunnage Mélange at James Island by Hibbard *et al.* (1977). Stouge (1980b) also reported a fauna comprising *Periodon*, *Polonodus* and *Paroistodus horridus* from the Cutwell Group on Long Island and the correlative Western Arm Group on Limestone Island. This fauna correlates with the Table Point Formation and the upper part of the Stein Formation.

Middle Ordovician conodonts from northern Central Newfoundland were recorded from the Cobbs Arm Formation at New World Island (Bergström *et al.* 1974; Fåhræus & Hunter 1981) and from the Davidsville Group at Weir's Pond (Stouge 1980a). The two units contain a very similar fauna, which led Stouge (1980a) to conclude that the proposed Iapetus suture between the two localities (McKerrow & Cocks 1977) should be questioned.

The Elvdal Formation shares several species with the Newfoundland localities and the significance of this was discussed by Rasmussen & Stouge (1989). The Cobbs Arm Formation contain for example *Pygodus serra* (Hadding), *Pygodus anserinus* Lamont & Lindström, *Eoplacognathus lindstroemi* Hamar, *E. robustus* (Bergström), *Periodon aculeatus* Hadding, *Baltoniodus variabilis* (Bergström) and *B. prevariabilis* Fåhræus. The fauna of the Davidsville Group comprises the same species, except *P. anserinus* and *B. variabilis* are replaced by *P.* cf. *anserinus* and *B.* cf. *variabilis*.

It is concluded that the Davidsville Group correlates with the "Andersö Shale" interval somewhere between sample 99484 (containing *P. serra* and *Eoplacognathus reclinatus*) and the basal part of the "Biseriata Limestone" (with *P. anserinus*) at Andersön. The Cobbs Arm Formation correlates with the interval between sample 99484 and an unknown level above the "Biseriata Limestone" at Andersön (see also Bergström *et al.* 1974, table 10), whilst the upper part of the Cobbs Arm Formation correlates with the Elvdal Formation of eastern Norway (*P. anserinus-A. inaequalis* Subzone).

Hølonda, Upper Allochthon, central Norway

The Hølonda Limestone is part of the Lower Hovin Group from the Upper Allochthon Støren Nappe. The Hølonda Limestone was deposited around an island-arc "greenstone island" (Bruton & Bockelie 1980), located within the Iapetus Ocean. Bergström (1979) demonstrated that the Hølonda Limestone conodont fauna displays relations to the Whiterockian faunas of North America, especially the Table Head Group of western Newfoundland (see also Stouge 1984 and Rasmussen 1997).

Characteristic taxa include (with Bergström's original designations in brackets): *Histiodella holodentata* Ethington & Clark [= *Histiodella* cf. *serrata* Harris], *Plectodina?* sp. A Stouge 1984 [= *Plectodina?* sp.], *Parapanderodus elegans* Stouge [= "*Scolopodus*" sp.], *Paroistodus horridus* (Barnes & Poplawski) [= "*Cordylodus*" *horridus*] and *Ansella* cf. *sinuosa* (Stouge) [= *Belodella* sp. A Fåhræus 1970]. This assemblage indicates that the Hølonda Limestone correlates with a level within the *B. medius– H. holodentata* Zone, i.e. lower Llanvirn.

North America, Utah, Ibex

The conodont fauna of the Pogonip Group and the adjoining limestone units at Ibex, Utah, was described by Ethington & Clark (1982) (Fig. 28). The studied interval includes the Ross-Hintze shelly fossil zones B–O.

Although the fauna belongs to the Midcontinent Faunal Province, it has about nine taxa in common with the Baltoscandian margin, which provide important stratigraphical ties between this province and Baltoscandia.

The occurrence of *Acodus deltatus*, *Fahraeusodus* aff. *marathonensis* (= "*Microzarkodina*" *marathonensis* of Ethington & Clark) and *Paroistodus parallelus* within the lower and middle part of the Fillmore Formation (the *Acodus deltatus– Macerodus dianae* Interval) indicates that most of this interval (except the *Protopanderodus*-bearing upper part), correlates with the *Paroistodus proteus* Zone.

Protopanderodus spp. and *Paroistodus parallelus* co-occur in the uppermost *Jumodontus gananda–?Reutterodus andinus* Interval within the upper part of Fillmore Formation and the lowest part of the succeeding Wahwah Formation (Ross-Hintze zones H–J). This suggests a correlation with a level (partly?) within the *Oepikodus evae* Zone at Andersön. Moreover, *Jumodontus gananda* was recorded from the *Microzarkodina* sp. A Zone of Öland (Bagnoli & Stouge 1997), which correlates with the upper *O. evae* Zone as defined in this paper. ?*Reutterodus andinus*, which is one of the zonal species at Ibex, is restricted to the *O. evae* Zone within the Cow Head Group (Stouge & Bagnoli 1988).

The uppermost part of the succeeding Wahwah Limestone (the upper part of the *Microzarkodina flabellum – Tripodus laevis* Interval and the Ross-Hintze Zone K) contains *Microzarkodina* sp. (= *M. flabellum sensu* Ethington & Clark), *Protopanderodus* spp. and *Protoprioniodus* spp. These genera also occur in the Herram Member of the Steinsodden Formation, and a correlation with the *Microzarkodina flabellum – D. forceps* Zone is possible. However, the *Microzarkodina* P-elements figured by Ethington & Clark (1982, pl. 5, figs. 21, 26) are characterised by a low angle between the anterior denticle and the cusp, which characterises *Microzarkodina parva* Lindström. Accordingly, the upper part of the *M. flabellum – T. laevis* Interval may instead correlate with a level within the *P. rectus – M. parva* Zone.

The last tie is marked by the appearance of *Histiodella holodentata* in the upper part of the Lehman Formation (*Paraprioniodus costatus – Chosonodina rigbyi – Histiodella holodentata* Interval; Ross-Hintze Zone N), which strongly suggests correlation with the *B. medius – H. holodentata* Zone. Judging from the synonymy included by Ethington & Clark (1982), it is possible that they included specimens, which is now known to be part of *Histiodella kristinae* Stouge within *H. holodentata*. This means that their *H. holodentata*-bearing interval *may* extend into the *Eoplacognathus suecicus* Zone, as *H. kristinae* appears close to the base of this zone in the sections studied herein.

In conclusion, the following ties between the zones and intervals of Ethington & Clark and the Baltoscandian margin zones are recognised: 1) The "*Scolopodus*" *quadraplicatus* – aff. *Scolopodus rex* Interval and *Acodus deltatus – Macerodus dianae* Interval correlate with the *Paroistodus proteus* Zone; 2) The *Jumodontus gananda – ?Reutterodus andinus* Interval correlates – at least in part – with the *Oepikodus evae* Zone; 3) The *Microzarkodina flabellum – Tripodus laevis* Interval correlates with a level within the interval from the *Microzarkodina flabellum – D. forceps* Zone to the *P. rectus – M. parva* Zone at the Baltoscandian margin, the latter zone being the most probable. It was suggested by Finney & Ethington (1992, p. 166) that the *M. flabellum – T. laevis* Interval correlates with an interval within the North American Midcontinent Zone of *Histiodella altifrons*, and that the base of this corresponds to the base of the Whiterock stage. 4) The *Paraprioniodus costatus – Chosonodina rigbyi – Histiodella holodentata* Interval correlates with the *B. medius – H. holodentata* Zone of the Baltoscandian margin and possibly also with the overlying interval from the *M. ozarkodella* Zone to the basal *E. suecicus* Zone.

Greenland

Early Ordovician conodonts from the Cape Weber Formation, Ella Ø and the Wandel Valley Formation, Kronprins Christian Land and Peary Land were described by Smith (1991), while Stouge *et al.* (1985) described Tremadoc conodonts from the Cape Clay Formation of Washington Land, western North Greenland. The Greenland platform fauna is a typical Midcontinent Fauna Province association, which has very few species in common with the Baltoscandian margin. However, the two first-mentioned sections studied by Smith (1991) are characterised by an upper part that contains, for example, *Protopanderodus*. The appearance of this genus together with *Fahraeusodus* aff. *marathonensis* (= *F. marathonensis* of Smith) at a level within the *Oepikodus communis* Biozone indicate correlation with the *P. elegans* or *Oepikodus evae* Zone. The interval just below the simultaneous occurrence of *Protopanderodus* and *F.* aff. *marathonensis* thus may correlate with the *P. proteus* Zone.

The upper part of the Cape Weber Formation at Ella Ø contains, for example, ?*Reutterodus andinus* Serpagli and *Toxotodus carlae* (Repetski), which are typical of the *O. evae* Zone at Cow Head (Stouge & Bagnoli 1988), and thus indirectly favour correlation with the *O. evae* Zone at Andersön.

The Cape Weber Formation is succeeded by the Narwhale Sound Formation at Ella Ø. The appearance of *Pteracontiodus cryptodens* (Mound) 95 m above the base, indicates correlation with the lower part of the Kanosh Shale of Ibex (the *P. cryptodens – H. altifrons – M. auritus* Interval of Ethington & Clark (1982). The interval is above the appearance of "*Microzarkodina flabellum*" (probably *M. parva*) and below the appearance of *Histiodella holodentata* at Ibex (Ethington & Clark 1982). Accordingly, correlation with the interval from the *P. rectus – M. parva* Zone to the *B. norrlandicus – D. stougei* Zone of Baltoscandia should be expected for this part of the Cape Weber Formation.

The upper half of the Narwhale Sound Formation is characterised by *Histiodella holodentata*, *Chosonodina rigbyi* Ethington & Clark and *Multioistodus* (Smith 1982) which – according to the correlation of the Ibex section discussed above – indicate correlation with a level within the *B. medius – H. holodentata* Zone – *E. suecicus* Zone interval.

The stratigraphically youngest tie between the East Greenland successions and the sections studied herein, is the occurrence of *Pygodus* cf. *anserinus* (Lamont & Lindström) within the upper part of the Heim Bjerge Formation at C. H. Ostenfeld Nunatak, East Greenland. This species indicate correlation with the *P. anserinus* Zone of the Elvdal Formation and the equivalent "Biseriata Limestone" on the Baltoscandian margin.

Argentina

Limestones of the San Juan Formation from the Precordillera of the San Juan Province, western Argentina contain a rich conodont succession that is similar to that of Laurentia, but like the Ibex area, it also shows some degree of similarity with Baltoscandia. The lower Ordovician succession was described by Serpagli (1974) and Lehnert (1995a), while Hünicken and his co-workers and Lehnert (1995a) have described the faunas of the uppermost part of the San Juan Formation and the limestone beds within the succeeding Los Azules Formation (Hünicken & Ortega 1987) and the laterally equivalent Gualcamayo Formation (Albanesi *et al.* 1995).

Serpagli (1974) divided the investigated part of the San Juan Formation into five intervals or faunas. The two lowermost samples were included within Fauna A, which contains for example *Paracordylodus gracilis* Lindström, *Drepanodus arcuatus* Pander, *Paroistodus proteus* (Lindström) (included in *P. parallelus* (Pander) by Serpagli), *Protopanderodus leonardii* Serpagli and *Drepanoistodus forceps* (Lindström). *Periodon selenopsis* (Serpagli), which is restricted to the upper sample, is typical of the *P. elegans* and *O. evae* zones at Cow Head (Stouge & Bagnoli 1988) and the Baltoscandian margin. This suggests that the lower "Fauna A" sample correlates with the *P. proteus* Zone and the upper sample with the *P. elegans* Zone. The succeeding "Fauna B" is characterised by the occurrence of *Oepikodus evae* (Lindström) and *Acodus? gladiatus* Lindström, which correlates with the *Oepikodus evae* Zone. The disappearance of *Drepanoistodus forceps* from the base of "Fauna E" *may* indicate correlation with the upper boundary of the *M. flabellum – D. forceps* Zone, but the absence might also be due to an ecological change, because no *Drepanoistodus* specimens were recorded above this level. The occurrence of *Triangulodus brevibasis* Van Wamel and *D. forceps* in the lower-middle part of "Fauna D" suggests correlation with a level within the *M. flabellum – D. forceps* Zone. It is more difficult to correlate the sparse fauna of "Fauna E" with the faunal succession studied herein. The presence of *Panderodus* sp., however, points to a correlation with the *E. suecicus* Zone or younger. This correlation is supported by the presence of *Eoplacognathus* cf. *suecicus* Bergström (= *E. suecicus sensu* Hünicken & Ortega 1987) together with *Polonodus tablepointensis* Stouge in the uppermost part of the San Juan Formation at San José de Jáchal in the northern part of the San Juan Province (Hünicken & Ortega 1987). There, the overlying Los Azules Formation consists of a c. 100 m thick, *Pygodus serra*-bearing, shale sequence, succeeded by c. 15 m thick calcareous unit that contains *Nemagraptus gracilis* (Hall).

Albanesi *et al.* (1995) summarised the conodont biostratigraphy of the Argentine Precordillera formations, and showed that the boundary between the calcareous San Juan Formation and the overlying shaly Gualcamayo Formation is highly diachronous varying from Middle Arenig in the northern Precordillera to Lower Llanvirn in the south (Albanesi & Barnes 1996). It has been suggested that the sedimentological transition in most parts of the Central Precordillera is a result of crustal extension, which in many places caused erosion of the top of the San Juan Formation before the subsequent drowning of the platform (Lehnert 1995b).

Lehnert (1995a) showed that the conodont faunas from the platform facies of the eastern part of the Argentine Precordillera are quite similar to those of North America. In his comprehensive work, he established 8 biozones for the Tremadoc – lower Llanvirn interval, which were accurately correlated with the North American faunal succession.

Discussion of the mixed Herram fauna

The most obvious explanation of the mixing of the "older species" and the "younger species" in the Herram section is reworking. The "older species" perhaps were transported to the Herram locality from positions on the more shallow part of the platform, e.g. by gravity flows. With this in mind, it should be noted that Heath & Owen (1991) reported channel deposits interpreted as debris flows originating from a carbonate platform to the east from the Upper Ordovician Kalvsjøen Formation of Hadeland in the Oslo Region. However, in the Herram section no obvious sedimentological evidence of reworking (conglomerates, crossbedding etc.) has been demonstrated. Similarly, the conodont elements lack physical indication of reworking although *Stolodus stola* elements typically show a little more light reddish-brown colour than other taxa.

Alternatively, the explanation for the mixed aspect of the fauna is that the "older species" survived longer near the shelf edge than in the rest of Baltoscandia (Kohut 1972). It is also possible that *Paracordylodus gracilis*, *Stolodus stola* and *Oepikodus evae* migrated from the more proximal parts of the platform to the shelf edge during the regression in the upper part of the *O. evae* Zone. In that case, one should expect a diachronous disappearance of the three species across the platform. Löfgren (1993b, Fig. 8) extended the vertical ranges of *S. stola* and *O. evae* to a level within the lower part of the *B. triangularis* Zone with a punctuated line. The spatial distribution of these species is still cryptic, because of their absence in the Volkhovian successions of Jämtland (Löfgren 1978; this paper). At the present, there is no obvious explanation on the unusual species association within the Herram Member.

Correlation with the local graptolite and trilobite zonations

Only very few macrofossils have been described from the Stein and Elvdal formations due to poor preservation caused by pressure-dissolution. However, graptolites and trilobites are well known from the surrounding shale units.

At Herram, Ringsaker, the Tøyen Formation, which separates the Tremadoc Bjørkåsholmen Formation and the Arenig–Llanvirn Stein Formation, contains graptolites indicative of the Billingen *Phyllograptus densus* and *Phyllograptus angustifolius elongata* graptolite zones (Skjeseth 1952) (Fig. 29). The succeeding Herram Member of the Stein Formation correlates with the Volkhovian *Didymograptus hirundo* Zone (Kristina Lindholm, pers. comm. 1992).

Tjernvik (1956) studied the Tøyen Formation at the Andersön-A locality. He divided the formation into five units: B (Zone of *Megistaspis armata*), C (Dark shale, about 2 m thick, without fossils), D: (Zone of *Megistaspis planilimbata* and *Tetragraptus phyllograptoides*), E (Zone of *Didymograptus balticus*) and F (Intercalating limestone beds and marly shale, where he noted "No fossils encountered, Billingen group?"). According to his lithological log (Tjernvik 1956, Fig. 25), Zone D correlates with the *P. proteus* Zone, Zone E with the *P. elegans* Zone and Zone F with the *O. evae* Zone and the basal part of the *M. flabellum – D. forceps* Zone.

The Elnes Formation and the lateral equivalent "lower Andersö shale", which succeeds the Stein Formation at Andersön, contain abundant trilobite and graptolite faunas. *Ogygiocaris dilatata* (Brünnich), *Didymograptus geminus* (Hisinger) and *Pseudoclimacograptus scharenbergi* (Lapworth) were recorded from the Elnes Formation ("4a") at Ringsaker (Holtedahl 1909; Strand 1929; Skjeseth 1963) whereas *Ogygiocaris* sp. and *Hustedograptus teretiusculus* (Hisinger) have been reported from several localities west of Lake Mjøsa (Münster 1900). A. Bjørlykke (1979) recorded *Ogygiocaris* sp. and *Diplograptus* sp. from Snartumelva, eastern Snertingdal, and Flåmyra (Aust-Torpa). None of this material was figured. Holtedahl (1909, p. 13) restudied the fauna from the Elnes Formation ("4a") at Stokbækken, Vest-Torpa, collected by Münster (1900) and recorded *O. sarsi* Angelin, *Trinucleus foveolatus* Angelin, *Cheirurus* sp., *Climacograptus* sp. and *Diplograptus* sp.

The Elnes Formation from Ringsaker and Snertingdal-Torpa shares the graptolites *D. geminus*, *P. scharenbergi*, *H. teretiusculus* and *Climacograptus* spp.

with the Elnes Formation at Slemmestad, Oslo-Asker (Berry 1964), and thus correlates with the *Didymograptus murchisoni* Zone. The occurrence of the trilobite species *Trinucleus foveolatus* within the Elnes Formation at Torpa, however, may indicate that parts of the formation correlates with the lower part of the *H. teretiusculus* Zone (Størmer 1953, Fig. 16).

Tjernvik (1956) listed *Megistaspis* sp. from about 0.5 m above the base of the formation at the Andersön-A section, while Tjernvik & Johansson (1980 p. 197) recorded *Megistaspis (M.) lata* (Törnquist) from the lowermost part of the Stein Formation. *Asaphus expansus* (Linné) was recorded from a level c. 5 m above the base of the Stein Formation by Karis (1982), and *Paraceraurus exsul* (Beyrich) and *Pseudomegalaspis patagiata* (Törnquist) from the uppermost 2 m of the formation. Based on this, Karis (1982) correlated the top of the formation with the Lasnamägian Folkeslunda Limestone (upper Llanvirn). However, this conflicts with the conodont fauna, which correlates with the *Eoplacognathus suecicus* Zone (middle Llanvirn). Hadding (1912, tab. 7a, loc. 1) reported the trilobites *Pseudasaphus tecticaudatus* (Steinhart) and *Trinucleus coscinorhinus* Angelin from the Stein Formation (possibly the uppermost part) at the locality Andersön-B.

The Stein Formation is succeeded by the lower "Andersö shale", which is about 6 m thick at Andersön-B, though the contact is faulted. The lower "Andersö shale" of Andersön, Jämtland, is equivalent to Elnes Formation of Norway, and should probably be referred to this. It is followed by the 1.5 m thick "Biseriata Limestone" (named after the trilobite *Telephina biseriata* (Asklund)) (Fig. 21), which is again overlain by more than 5 metres of "Andersö shale" (Hadding 1912, 1913; Karis 1982). The lower part of the lower "Andersö shale" unit contains *Ogygiocaris sarsi*, *Pseudoclimacograptus scharenbergi*, "*Climacograptus*" *putillus* (Hall) (= "*Climacograptus*" *haddingi*) (Hadding 1912) indicating correlation with the basal part of the *H. teretiusculus* Zone. The succeeding part of the lower "Andersö shale" consists of shales with interbedded limestone beds that yield e.g. *O. sarsi*, *Pseudomegalaspis patagiata*, *Botryoides bronni* (Sars) (Karis 1982) and correlates with a higher level within the *H. teretiusculus* Zone. This is in agreement with the present observations of *Pygodus serra* (Hadding) and *Eoplacognathus reclinatus* (Fåhræus) within the latter unit (sample 99484) according to the graptolite-conodont ties reported by Bergström (1973, 1986).

The overlying "Biseriata Limestone" is quite fossiliferous in places (Bergström, pers. comm. 1997), which is further supported by the extensive fauna list from the *H. teretiusculus* Zone reported by Hadding (1913). The

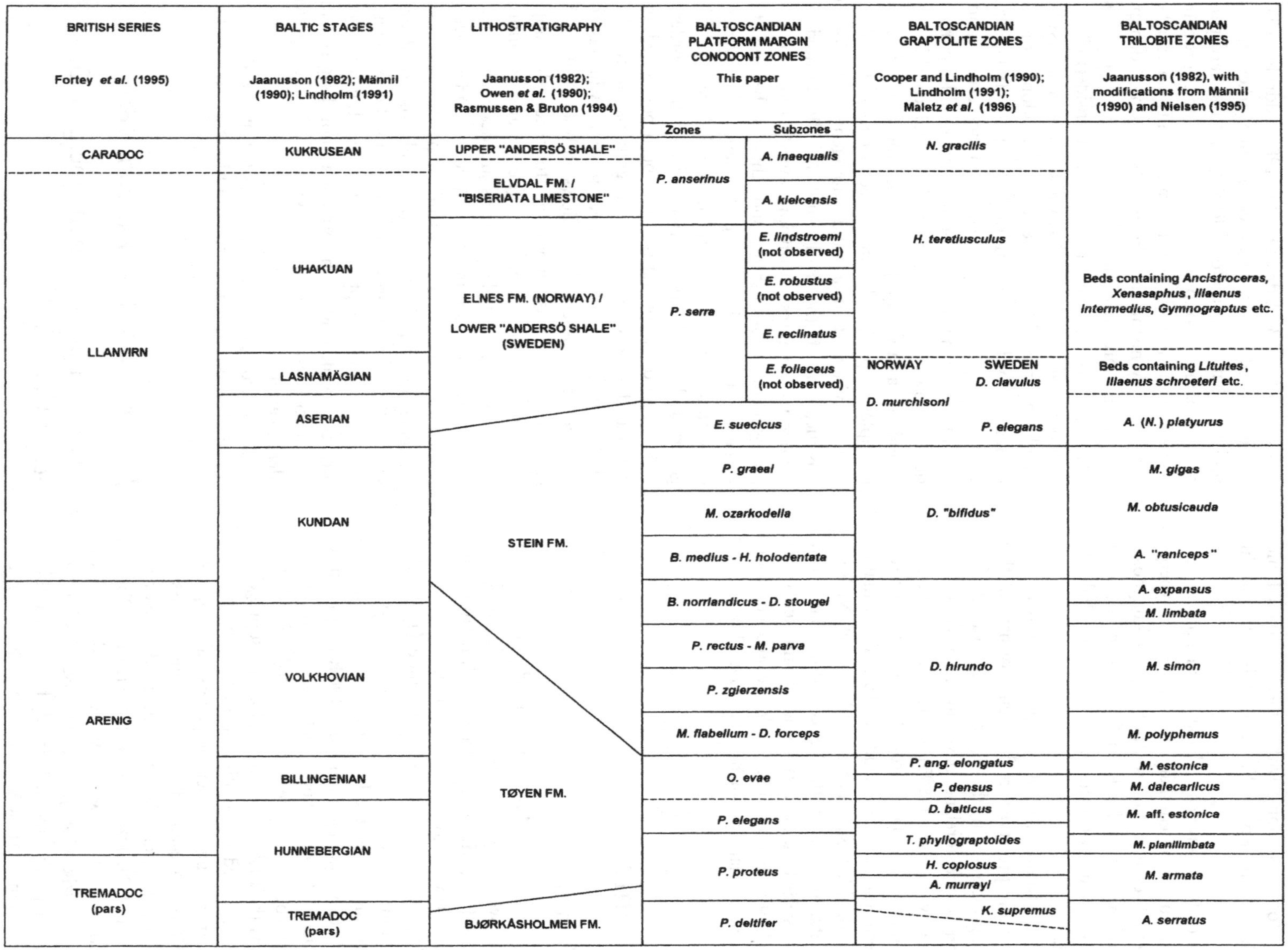

BRITISH SERIES Fortey *et al.* (1995)	BALTIC STAGES Jaanusson (1982); Männil (1990); Lindholm (1991)	LITHOSTRATIGRAPHY Jaanusson (1982); Owen *et al.* (1990); Rasmussen & Bruton (1994)	BALTOSCANDIAN PLATFORM MARGIN CONODONT ZONES This paper — Zones	Subzones	BALTOSCANDIAN GRAPTOLITE ZONES Cooper and Lindholm (1990); Lindholm (1991); Maletz *et al.* (1996)	BALTOSCANDIAN TRILOBITE ZONES Jaanusson (1982), with modifications from Männil (1990) and Nielsen (1995)
CARADOC	KUKRUSEAN	UPPER "ANDERSÖ SHALE"	*P. anserinus*	*A. inaequalis*	*N. gracilis*	
	UHAKUAN	ELVDAL FM. / "BISERIATA LIMESTONE"		*A. kielcensis*		Beds containing *Ancistroceras, Xenasaphus, Illaenus intermedius, Gymniograptus* etc.
LLANVIRN		ELNES FM. (NORWAY) / LOWER "ANDERSÖ SHALE" (SWEDEN)	*P. serra*	*E. lindstroemi* (not observed)	*H. teretiusculus*	
				E. robustus (not observed)		
				E. reclinatus		
	LASNAMÄGIAN			*E. foliaceus* (not observed)	NORWAY SWEDEN *D. clavulus*	Beds containing *Lituites, Illaenus schroeteri* etc.
	ASERIAN		*E. suecicus*		*D. murchisoni* *P. elegans*	*A. (N.) platyurus*
	KUNDAN	STEIN FM.	*P. graeai*			*M. gigas*
			M. ozarkodella		*D. "bifidus"*	*M. obtusicauda*
			B. medius - H. holodentata			*A. "raniceps"*
			B. norrlandicus - D. stougei		*D. hirundo*	*A. expansus*
			P. rectus - M. parva			*M. limbata*
	VOLKHOVIAN		*P. zgierzensis*			*M. simon*
			M. flabellum - D. forceps			*M. polyphemus*
ARENIG	BILLINGENIAN	TØYEN FM.	*O. evae*		*P. ang. elongatus* *P. densus* *D. balticus*	*M. estonica* *M. dalecarlicus* *M. aff. estonica*
	HUNNEBERGIAN		*P. elegans*		*T. phyllograptoides*	*M. planilimbata*
			P. proteus		*H. copiosus* *A. murrayi*	*M. armata*
TREMADOC (pars)	TREMADOC (pars)	BJØRKÅSHOLMEN FM.	*P. deltifer*		*K. supremus*	*A. serratus*

Fig. 29. Correlation of the conodont zones with Baltoscandian graptolite and trilobite zones.

succeeding upper "Andersö shale" comprises *Nema-graptus gracilis* Hall, indicative of the *N. gracilis* Zone. This corresponds well with the occurrence of *P. anserinus* Zone conodonts within the "Biseriata Limestone" (Bergström, Riva & Kay 1974, table 10; and this study).

The laterally equivalent Elvdal Formation is succeeded by an unnamed shale unit at both Høyberget and Engerdal, eastern Norway, but no zone-diagnostic fossils have been recorded from this unit so far, but it probably correlates with the upper "Andersö shale" of Jämtland.

Although trilobites and other macrofossils are generally badly preserved in the Stein Formation, they are well preserved in the laterally equivalent Huk Formation of the foreland basin in the Oslo Graben area, Norway. Nielsen (1995) made a comprehensive study of the trilobite fauna from the Arenig succession of southern Scandinavia, including the Huk Formation at Djuptrekkodden, Slemmestad (Oslo-Asker). Nielsen's (1995) data, together with the conodont data from exactly the same beds at Slemmestad (Rasmussen 1991), firmly links the Baltoscandian trilobite and conodont zones of the foreland basin together (Fig. 29).

Taxonomy

Introduction

The first attempts to create a classification of conodonts were based on form taxonomic considerations (Hass 1962, with earlier references), meaning that each element was given a genus and species name. Although it was shown already by Hinde (1879) that one conodont species may comprise different elements, it was not until after the key-papers in the middle of the 1960's (for example Lindström 1964; Bergström & Sweet 1966; Webers 1966) that the multielement-taxonomy became generally accepted. Lindström (1970) was the first to propose a biologic, suprageneric classification based on multi-element taxonomy. Since then, an increasing number of papers (although still relatively few) have aimed at a suprageneric classification. Ordovician examples were included by Sweet & Bergström (1972), Dzik (1976), the Treatise (Robison 1981); Stouge (1984); Sweet (1988); Aldridge & Smith (1993) and Dzik (1994). The Treatise classification, however, was partly outdated already before its publication (the main part was written before 1976) and contained several errors according to Fåhræus (1984).

Many of the above-mentioned papers refer the conodonts to the phylum Conodonta Eichenberg 1930. The Treatise (Robison 1981) included one class,

two orders, ten superfamilies and several families in the phylum. Since the discovery of the "conodont animal" from the Carboniferous Granton beds of Scotland (e.g. Briggs *et al.* 1983; Aldridge *et al.* 1986; Aldridge 1987; Aldridge *et al.* 1993), many conodont workers find it probable that the conodont elements are the hard parts of a separate group of jawless craniates, probably related to the hagfish. A vertebrate affinity was further supported by histological investigations, which demonstrated the presence of cellular bone in conodont elements (Sansom *et al.* 1992). Considering the hypothesis of Sansom *et al.* is valid, conodonts possibly should be regarded as the class Conodonta of the phylum Chordata (e.g. Aldridge & Smith 1993; Dzik 1994). Accordingly, the suprageneric classification should be changed dramatically. It should be mentioned, however, that some authors still consider the conodonts as a group distinct from the vertebrates, for example a separate phylum (Sweet 1988).

It is evident that a valid suprageneric, biological classification of the Early and Middle Ordovician conodonts requires a major investigation of the phylogenetic relationships and this is beyond the scope of this work. As a consequence, suprageneric classification has not been employed here, and the genera and species are described in alphabetic order.

The conodont apparatuses have been positioned and designated according to the suggestions of Robison (1981) and Sweet (1988) where possible, whereas the morphological description follows Robison (1981). In cases where it was impossible to place the single elements in the apparatus (i.e. many coniform apparatuses), the single elements are designated using the assumed correct position in form-element genus name with the suffix "-form". The use of open nomenclature follows the recommendations of Bengtson (1988). The abbreviations used in the synonymy lists are as follows: pt. = *in partim* (partly); aff. = affinity; cf. = confer; non = the present author disagree with the species and/or genus assignment; ? = the species identification is uncertain.

All figured specimens are deposited at the Palaeontological Museum, University of Oslo, Norway (PMO).

Phylum Chordata Bateson, 1886
Class Conodonta Eichenberg, 1930 *sensu* Clark 1981

Genus *Acodus* Pander, 1856

Type species. – Acodus erectus Pander, 1856

Remarks. – The multielement genus *Acodus* has been interpreted as comprising prioniodontiform, oistodontiform and ramiform elements (referred to P, M, S in this study) (Lindström 1977).

Acodus deltatus Lindström, 1955

Pl. 1: 1

Synonymy. –

1955 *Acodus deltatus* n. sp. – Lindström, p. 544; Pl. 3: 30.

pt. 1977 *Acodus deltatus deltatus* Lindström – Lindström, pp. 7–8, Acodus- plate 2: 10–13 (*non* Figs. 8, 9 = *Nordiora? crassulus* [Lindström]).

1988 *Baltoniodus? deltatus* (Lindström) – Bagnoli *et al.*, pp. 208–209; Pl. 38: 8–14 (*cum. syn.*).

1993a *Acodus deltatus deltatus* Lindström – Löfgren, Fig. 8c–d.

1993b *Acodus deltatus deltatus* Lindström – Löfgren, Fig. 5q–v.

1995a *Acodus? deltatus* Lindström – Lehnert, p. 68; Pl. 3: 1–5, 8.

Remarks. – The interpretation of the multielement species *Acodus deltatus* of Bagnoli *et al.* (1988) is followed herein. The individual elements were described by Lindström (1955a).

Occurrence. – Andersön-A, *P. proteus – O. evae* zone.

Material. – One Pa and one Sb element.

Genus *Amorphognathus* Branson & Mehl, 1933

Type species. – *Amorphognathus ordovicicus* Branson & Mehl, 1933

Amorphognathus? sp.

Pl. 1: 3

Description. – The element is a broken, pastiniscaphate Pa element with a slender, blade-like anterior process, a wide posterior process and a bilobate postero-lateral process. All processes are denticulated. The postero-lateral process is clearly shorter than the other processes. The basal cavity is wide and extends to the margin of the processes.

Remarks. – The specimen resembles *Amorphognathus inaequalis* Rhodes 1953 but because associated element-types are lacking and the specimen is broken, no generic or specific assignment has been made.

Occurrence. – Andersön-B, *P. serra – E. reclinatus* Subzone.

Material. – 1 Pa element

Genus *Ansella* Fåhræus & Hunter, 1985

Type species. – *Belodella jemtlandica* Löfgren, 1978

Remarks. – The *Ansella* apparatus was reconstructed by Löfgren (1978) based on the Early – Middle Ordovician species *Ansella jemtlandica* (Löfgren). It comprises P, M, Sa, Sb and Sc elements.

A. jemtlandica originally was referred to the genus *Belodella* Ethington by Löfgren (1978), but it was later transferred to the new genus *Ansella* by Fåhræus & Hunter (1985) because of the lack of M elements in *Belodella*.

Ansella jemtlandica (Löfgren, 1978)

Pl. 1: 4–9

Synonymy. –

Multielement

1978 *Belodella jemtlandica* n. sp. – Löfgren, p. 46; Pl. 15: 1–8; Text-fig. 24 A–D.

? 1981 *Belodella jemtlandica* Löfgren – Cooper, Pl. 26: 14.

? 1983 *Belodella jemtlandica* Löfgren – An *et al.*, Pl. 25: 8–12.

pt. 1984 *Belodella jemtlandica* Löfgren – Stouge, p. 60; Pl. 6: 13–23, *non* Pl. 7: 1–4.

non 1985a *Ansella jemtlandica* (Löfgren) – Fåhræus & Hunter, p. 1173; Pl. 1: 4, 6–9; Text-fig. 1.

aff. 1985 *Belodella fenxiangensis* An, Du, Gao, Chen & Lee – An, Du & Gao, Pl. 15: 13–20.

non 1989 *Ansella jemtlandica* (Löfgren) – Bauer, p. 98; Pl. 4: 4, 6, 10.

1991 *Ansella jemtlandica* (Löfgren) – Rasmussen, p. 273; Fig. 5L.

non 1995a *Ansella jemtlandica* (Löfgren) *sensu lato* – Lehnert, pp. 70–71; Pl. 9: 1, 2, 5; Pl. 10: 8; Pl. 12: 1; Pl. 13: 1).

1998b *Ansella jemtlandica* (Löfgren) – Zhang, pp. 50; Pl. 1: 5–9.

Comments to the synonymy list. – The elements illustrated by Stouge (1984, Pl. 7: 1–4), Fåhræus &

PLATE I

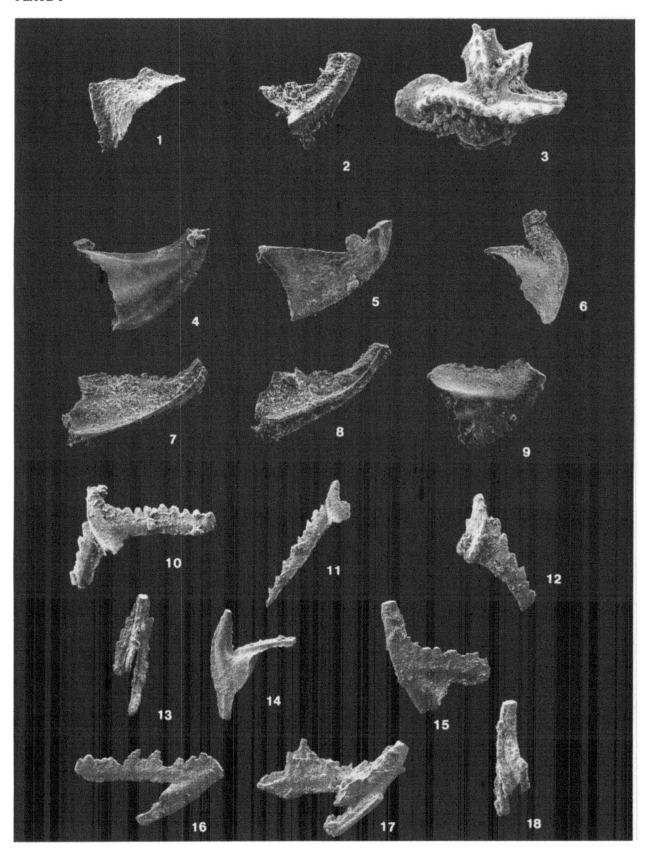

Hunter (1985a) and Bauer (1989) have relatively coarse denticles compared to the hair-like ones in *A. jemtlandica* and do probably not belong to this species. *Ansella fenxiangensis* (An, Du, Gao, Chen & Lee) pictured by An *et al.* (1985) shows the same fine denticulation on the S elements typically seen in *A. jemtlandica*, but is characterised by an atypical prolongation of the anterobasal part of the M element. The specimens depicted by Lehnert (1995a) are more coarsely denticulated than typical *A. jemtlandica* and co-occur with Sc and M elements of probable *Ansella nevadensis*.

Remarks. – *Ansella jemtlandica* was described in detail by Löfgren (1978, p. 46), but she did not distinguish between the Sa and Sb elements. The Sa element was described by Löfgren (1978) as the "denticulated triangular element". The Sb element resembles this except that the inner side is weakly compressed resulting in an asymmetrical outline.

Stouge (1984, p. 60) described two types of P elements from a closely related *Ansella* species and designated them "undenticulated biconvex element" and "undenticulated plano-convex element" respectively. Variations in the shape of the P elements also occur in the present material, but as the morphological transitions are gradual all variations have been referred to one P element type.

Occurrence. – Andersön-A, -B, -C, Steinsodden, Røste, Jøronlia, Haugnes, Hestekinn. *B. medius - H. holodentata* Zone – *E. suecicus* Zone.

Plate I

1: *Acodus deltatus* Lindström.
1. P element, × 60. Sample 97799, Andersön-A. PMO 165.162.

2: *Lundodus gladiatus* (Lindström)
2. ?P element, × 80. Sample 97805, Andersön-A. PMO 165.163.

3: *Amorphognathus?* sp.
3. Pa element, × 65. Sample 99484, Andersön-B. PMO 165.164.

4–9: *Ansella jemtlandica* (Löfgren).
4. P element, × 90. Sample 69662, Steinsodden. PMO 165.165.
5. P element, × 120. Sample 69662, Steinsodden. PMO 165.166.
6. M element, × 120. Sample 69662, Steinsodden. PMO 165.167.
7. Sa element, × 120. Sample 69662, Steinsodden. PMO 165.168.
8. Sb element, × 140. Sample 69662, Steinsodden. PMO 165.169.
9. Sc element, × 135. Sample 69662, Steinsodden. PMO 165.170.

10–18: *Baltoniodus medius* (Dzik).
10. Pa element, × 75. Sample 99479, Andersön-B. PMO 165.171.
11. Pa element, × 100. Sample 99473, Andersön-B. PMO 165.172.
12. Pb element, × 65. Sample 99473, Andersön-B. PMO 165.173.
13. Sa element, × 105. Sample 99473, Andersön-B. PMO 165.174.
14. M element, × 70. Sample 99473, Andersön-B. PMO 165.175.
15. Sc element, × 100. Sample 99473, Andersön-B. PMO 165.176.
16. Sc element, × 100. Sample 99473, Andersön-B. PMO 165.177.
17. Sd element, × 100. Sample 99473, Andersön-B. PMO 165.178.
18. Sd element, × 100. Sample 99473, Andersön-B. PMO 165.179.

Material. – 175 P, 62 M, 46 Sa, 75 Sb, 45 Sc.

Genus *Baltoniodus* Lindström, 1971

Type species. – *Prioniodus navis* Lindström, 1955

Remarks. – The skeletal apparatus is septimembrate (Dzik 1976), and comprises Pa, Pb, M, Sa, Sb, Sc and Sd elements.

Baltoniodus medius (Dzik, 1976)

Pl. 1: 10–18

Synonymy. –

Multielement
1976   *Prioniodus alatus medius* ssp. n. – Dzik, p. 423; Pl. 42: 1; Text-fig. 23a–h, ?Text-fig. 23i–l.
1978   *Prioniodus (Baltoniodus) prevariabilis medius* Dzik – Löfgren, p. 86; Pl. 12: 27–36, Pl. 13: 1A, 1B, 3, 6A–6D (*cum. syn.*).
non 1984 *Baltoniodus? prevariabilis medius* (Dzik) – Stouge, p. 77; Pl. 15: 1–6.
1991   *Baltoniodus medius* (Dzik) – Rasmussen, p. 274; Fig. 5D–H, J, K.
1998b   *Baltoniodus medius* (Dzik) – Zhang, p. 53; Pl. 2: 9–15.

Comments to the synonymy list. – In the present authors opinion, the elements illustrated by Dzik (1976, Text-fig. 23i–l (only)) possibly do not belong to *Baltoniodus medius* because the outer lateral process-like costa of the Sb element is placed very close to the anterior process, and therefore is more advanced. The elements referred to *Baltoniodus? prevariabilis medius* (Dzik) by Stouge (1984) are part of *Polonodus* Dzik (Löfgren 1990).

Discussion. – The original diagnosis of *B. medius* was based on specimens from erratic samples and was very brief: "Lateral branch of keislognathiform element short, sharpened" (Dzik 1976, p. 423). Unfortunately, Dzik did not describe *B. medius*, but the illustrated specimens (Dzik, 1976; Pl. 42: 1, and Text-fig. 23a–h) are similar to the stratigraphically young specimens of *B. medius* in the present material. As some Sb elements in stratigraphically young specimens of *B. norrlandicus* (Löfgren) may be characterised by a short, sharp lateral process as well, the diagnosis is not adequate.

Description. – *Baltoniodus medius* was revised and fully described by Löfgren (1978, pp. 86–87). In short, the species may be diagnosed as a *Baltoniodus* species with an

PLATE II

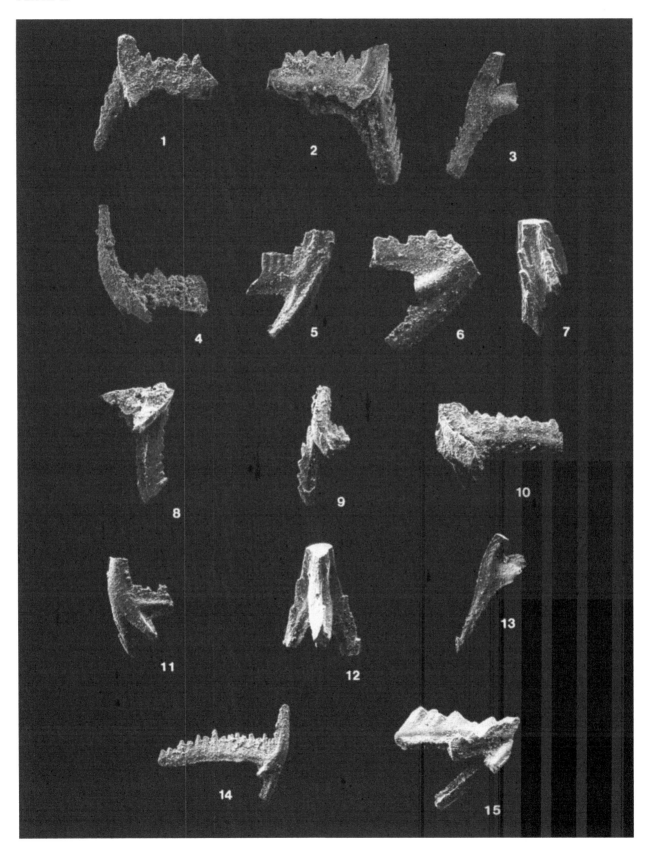

adenticulate anterior process on the Sb and Sc elements. The angle between the anterior and posterior processes is typically less than 50° on the Sb and Sc elements. The Sd element is asymmetrical to symmetrical with the processes situated 1 anterior - 2 lateral - 1 posterior. The M element is adenticulate (Löfgren 1978).

Remarks. – *B. medius* differs from *Baltoniodus* n. sp. A Stouge & Bagnoli 1990 by having a denticulated anterior process on the quadriramate Sd element, and from *B. clavatus* Stouge & Bagnoli 1990 by the existence of both symmetrical and asymmetrical Sd elements. Moreover, *B. medius* is characterised by an adenticulate anterior process in Sb and Sc elements, while the corresponding elements of *B. clavatus* occasionally may carry a distally situated denticle (Stouge & Bagnoli 1988, p. 13).

Occurrence. – Andersön-A, -B, -C, Steinsodden, Røste, Jøronlia, Haugnes, Hestekinn, Skogstad. *B. medius - H. holodentata* Zone – *E. suecicus* Zone.

Material. – 69 Pa, 141 Pb, 129 M, 44 Sa, 82 Sb, 76 Sc, 100 Sd.

Baltoniodus navis (Lindström, 1955)

Pl. 2: 1–7

Synonymy. –

Pa
pt. 1955a *Prioniodus navis* n. sp. – Lindström, p. 590; Pl. 5: 33 (only).

Multielement
1971 *Baltoniodus navis* (Lindström) – Lindström, p. 56; Pl. 1: 13, 19–23 (only), ?18.

Plate II
1–7: *Baltoniodus navis* (Lindström).
1. Pa element, ×85. Sample 99458, Andersön-A. PMO 165.180.
2. Pa element, ×70. Sample 69685, Steinsodden. PMO 165.181.
3. M element, ×60. Sample 69616, Steinsodden. PMO 165.182.
4. Sa element, ×80. Sample 69616, Steinsodden. PMO 165.183.
5. Sb element, ×80. Sample 99458, Andersön-A. PMO 165.184.
6. Sc element, ×90. Sample 99458, Andersön-A. PMO 165.185.
7. Sd element, ×100. Sample 99458, Andersön-A. PMO 165.186.
8–15: *Baltoniodus norrlandicus* (Löfgren).
8. Pa element, ×95. Sample 99461, Andersön-A. PMO 165.187.
9. Pb element, ×50. Sample 69622, Steinsodden. PMO 165.188.
10. Pb element, ×50. Sample 99461, Andersön-A. PMO 165.189.
11. Sc element, ×75. Sample 99461, Andersön-A. PMO 165.190.
12. Sa element, ×80. Sample 99461, Andersön-A. PMO 165.191.
13. M element, ×65. Sample 69622, Steinsodden. PMO 165.192.
14. Sc element, ×55. Sample 69622, Steinsodden. PMO 165.193.
15. Sd element, ×100. Sample 99461, Andersön-A. PMO 165.194.

1978 *Prioniodus (Baltoniodus) navis* Lindström – Löfgren, p. 83; Pl. 12: 8–16, Pl. 14: 1A–1B, 3A–3D.
1990 *Baltoniodus navis* (Lindström) – Stouge & Bagnoli, pp. 10–11; Pl. 1: 1–10 (*cum. syn.*).
1991 *Baltoniodus navis* (Lindström) – Rasmussen, p. 274; Fig. 5O–U.
1993b *Baltoniodus navis* (Lindström) – Löfgren, Fig. 6D.
1994 *Baltoniodus navis* (Lindström) – Löfgren, Fig. 8:23.
1995b *Baltoniodus navis* (Lindström) – Löfgren, Fig. 7: a–h.

Remarks. – The species were described fully by Löfgren (1978) and Stouge & Bagnoli (1990). The elements figured by Lindström (1955a, p. 590; Pl. 5: 31–32, 34–35 (only)) belong to *Baltoniodus triangularis* (Lindström) or *Trapezognathus quadrangulum* Lindström. The Pa element figured by Lindström (1971, Pl. 1: 18) belongs to *B. norrlandicus* (see next page).

Occurrence. – Andersön-A, -B, Herram, Steinsodden. *M. flabellum - D. forceps* Zone – *B. norrlandicus - D. stougei* Zone.

Material. – 24 Pa, 48 Pb, 45 M, 21 Sa, 34 Sb, 22 Sc, 43 Sd.

Baltoniodus norrlandicus (Löfgren, 1978)

Pl. 2: 8–15

Synonymy. –

Pa
pt. 1971 *Baltoniodus navis* (Lindström) – Lindström, p. 56; Pl. 1: 18 (only).

Multielement
1978 *Prioniodus (Baltoniodus) prevariabilis norrlandicus* n. ssp. – Löfgren, pp. 84–86; Pl. 10: 3A–E, Pl. 12: 17–26, Pl. 14: 2A–B (*cum. syn.*).
1983b *Baltoniodus navis* (Lindström) – Dzik, Text-fig. 7F–M.
1990 *Baltoniodus norrlandicus* (Löfgren) – Stouge & Bagnoli, pp. 11–12; Pl. 1: 11–20.
1991 *Baltoniodus norrlandicus* (Löfgren) – Rasmussen, p. 274; Fig. 5I, M–N.
? 1994 *Baltoniodus parvidentatus* (Sergeeva) – Dzik, pp. 80–82; Pl. 8: 8–14, textfigs. 13, 14a.
1995b *Baltoniodus norrlandicus* (Löfgren) – Löfgren, Fig. 9: a–i.
1997 *Baltoniodus norrlandicus* (Löfgren) – Bagnoli & Stouge, Pl. 2: 3–9.
cf. 1998b *Baltoniodus norrlandicus* (Löfgren) – Zhang, pp. 52–53; Pl. 1–8.

PLATE III

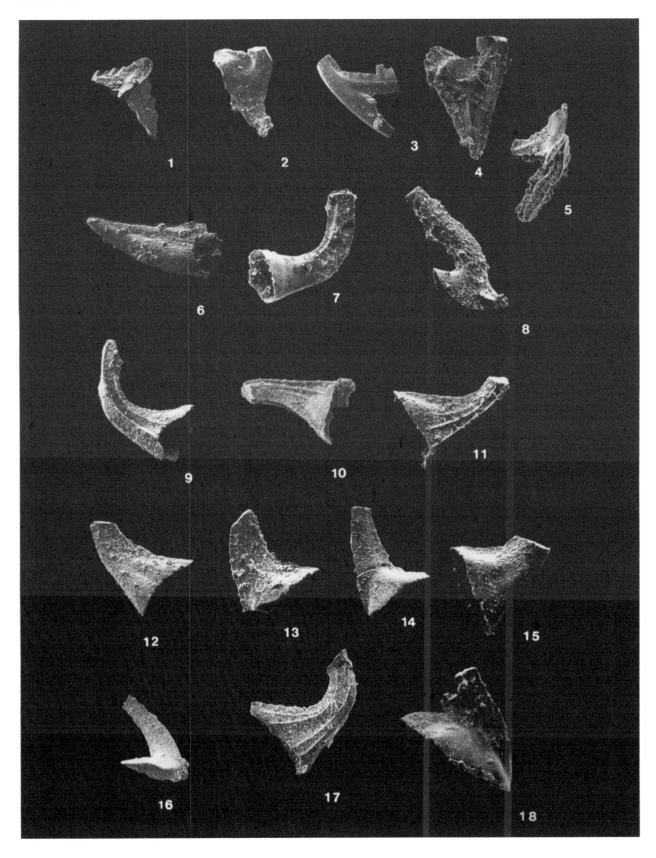

Remarks. – The species is characterised by a thickened posterior process on the Sb and Sc elements. The *B. norrlandicus* specimens at hand typically have an angle of 50 – 60° between the anterior and posterior process on the Sb and Sc elements, which is a little less than that in the specimens described by Löfgren (1978) (Sb elements 55 – 60° and Sc elements 55 – 70°). *B. norrlandicus* was described fully by Löfgren (1978). In contradiction to *B. norrlandicus*, the Sc element of *Baltoniodus parvidentatus* (Sergeeva) *sensu* Dzik (1994, Pl. 18: 14) sometimes shows alternating small and large denticles on the posterior process of the Sc element as in *B. medius*. Moreover, *B. parvidentatus* is typified by adenticulate anterior processes on the Sb and Sc elements, while the anterior process in *B. norrlandicus* commonly has distal denticles (Löfgren 1978). The *B. norrlandicus* specimens figured by Zhang (1998b) have an advanced, irregular denticulation on the S elements, which is typical of *Baltoniodus medius*. The angle between the anterior and posterior processes on the Sb and Sc elements, however, is within the range of *B. norrlandicus*.

Occurrence. – Andersön-A, -B, Steinsodden, Haugnes. *B. norrlandicus* - *D. stougei* Zone – lowermost *B. medius* - *H. holodentata* Zone.

Material. – 80 Pa, 158 Pb, 124 M, 55 Sa, 77 Sb, 106 Sc, 121 Sd.

Baltoniodus prevariabilis (Fåhræus, 1966)

Pl. 3: 1 – 3

Synonymy. –

Pa
1913 *Prioniodus alatus* n. sp. – Hadding, p. 32; Pl. 1: 9, 10 (*non Prioniodus alatus* Hinde 1879).

Pb
1966 *Prioniodus prevariabilis* n. sp. – Fåhræus, p. 29; Pl. 4: 5a – b.

M
1966 *Oistodus robustus* Bergström – Fåhræus, p. 24; Pl. 3: 3a – d; Text-fig. 2J.

Sa
1966 *Roundya inclinata* (Rhodes) – Fåhræus, p. 30; Pl. 3: 4.

Sc
1966 *Paracordylodus lindstroemi* Bergström – Fåhræus, p. 27; Pl. 3: 8.

Sd
1966 *Tetraprioniodus asymmetricus* Bergström – Fåhræus, p. 31; Pl. 3: 7a – b.

Multielement
1971 *Prioniodus prevariabilis* Fåhræus – Bergström, p. 146; Pl. 2: 1.
1978 *Prioniodus (Baltoniodus) prevariabilis prevariabi-*

Plate III

1–3: *Baltoniodus prevariabilis* (Fåhræus).
1. Pa element, × 95. Sample 99488, Andersön-B. PMO 165.195.
2. M elemen, × 80. Sample 99488, Andersön-B. PMO 165.196.
3. Sb element, × 95. Sample 99488, Andersön-B. PMO 165.197.

4: *Baltoniodus variabilis* (Bergström).
4. M element, × 75. Sample 99510, Høyberget. PMO 165.198.
5. Pb element, × 50. Sample 99510, Høyberget. PMO 165.199.

6: *Coelocerodontus?* sp.
6. Coniform element, × 125. Sample 97837, Andersön-A. PMO 165.200.

7–8: *Cornuodus longibasis* (Lindström).
7. Coniform element, × 125. Sample 69687, Steinsodden. PMO 165.201.
8. Coniform element, × 100. Sample 99471, Andersön-B. PMO 165.202.

9–15: *Costiconus costatus* (Dzik).
9. Costate element, × 95. Sample 99461, Andersön-B. PMO 165.203.
10. Costate element, × 90. Posterior view. Sample 99461, Andersön-B. PMO 165.204.
11. Same specimen as in Pl. 3:10. Lateral view, × 100.
12. Drepanodontiform element, × 70. Sample 99471, Andersön-B. PMO 165.205.
13. Oistodontiform element, × 105. Sample 99471, Andersön-B. PMO 165.206.
14. Oistodontiform element, × 75. Sample 99461, Andersön-B. PMO 165.207.
15. Drepanodontiform element, × 115. Sample 99461, Andersön-B. PMO 165.208.

16–18: *Costiconus ethingtoni* (Fåhræus).
16. Oistodontiform element, × 60. Sample 99479, Andersön-B. PMO 165.209.
17. Costate element, × 110. Sample 69675, Steinsodden. PMO 165.210.
18. Oistodontiform element, × 80. Sample 69675, Steinsodden. PMO 165.211.

lis Fåhræus – Löfgren, pp. 87 – 89; Pl. 12: 37 – 43 (*cum. syn.*).
1994   *Baltoniodus   prevariabilis*   (Fåhræus) – Dzik, pp. 82 – 84; Pl. 18: 17 – 22; Text-fig. 14b.
1995a *Baltoniodus   prevariabilis*   (Fåhræus) – Lehnert, pp. 73 – 74; Pl. 15: 8, 9, 11.
1998   *Baltoniodus prevariabilis* (Fåhræus) – Bednarczyk, Pl. 1: 9, 10, 12, 13.
1998b *Baltoniodus prevariabilis* (Fåhræus) – Zhang, p. 54; Pl. 3: 1 – 8.

Remarks. – The apparatus was reconstructed by Bergström (1971) and described in detail by Löfgren (1978).

Occurrence. – Andersön-A, *P. serra - E. reclinatus* Subzone – *P. anserinus-S. kielcensis* Subzone.

Material. – 10 Pa, 12 Pb, 9 M, 3 Sa, 11 Sb, 2 Sc, 4 Sd.

Baltoniodus variabilis (Bergström, 1962)

Pl. 3: 4 – 5

Synonymy. –

Pa
1960   *Prioniodus*? n. sp. 2 – Lindström, Fig. 8: 1.
pt. 1966 *Prioniodus alatus* Hadding – Hamar, p. 69; Pl. 5: 3; Text-fig. 6: 6 (only).

Pb
1960   *Prioniodus*? n. sp. 3 – Lindström, Fig. 8: 3.
1962   *Prioniodus variabilis* n. sp. – Bergström, pp. 51 – 53; Pl. 2: 1 – 7.
pt. 1966 *Prioniodus alatus* Hadding – Hamar, p. 69; Pl. 4: 6; Text-fig. 6: 5 (only).
pt. 1974 *Prioniodus variabilis* Bergström – Viira, Pl. 10: 16 – 17, ?18 (only).

M
1960 *Falodus* n. sp. 1 – Lindström, Fig. 8: 7.
1962 *Oistodus robustus* n. sp. – Bergström, p. 45; Pl. 3: 7 – 10; Text-fig. 3F.
1966 *Oistodus robustus* Bergström – Hamar, Pl. 1: 21.
1974 *Falodus robustus* (Bergström) – Viira, pp. 84 – 85; Pl. 10: 21 – 23; Text-fig. 100, 101.

Sa
1962 *Roundya* sp. – Bergström, p. 54; Pl. 5: 16.

Sb, Sc
1960 *Paracordylodus* n. sp. 2 – Lindström, Fig. 7: 15, Fig. 8: 6.
1962 *Paracordylodus lindstroemi* n. sp. – Bergström, pp. 50 – 51; Pl. 2: 8 – 12; Text-fig. 2C.

1966 *Paracordylodus lindstroemi* Bergström – Hamar, Pl. 7: 25.
1967 *Paracordylodus* n. sp. 2 Lindström 1960 – Viira, Fig. 3: 8.
1974 *Paracordylodus lindstroemi* Bergström – Viira, Pl. 10: 19, 20.

Sd
1960   *Tetraprioniodus* sp. – Lindström, Fig. 8: 4.
1962   *Tetraprioniodus asymmetricus* n. sp. – Bergström, pp. 55 – 56; Pl. 2: 15 – 17.
1966   *Tetraprioniodus asymmetricus* Bergström – Hamar, Pl. 5: 4.
pt. 1974 *Tetraprioniodus asymmetricus* Bergström – Viira, Pl. 10: 9 – 10 (only).

Multielement
1971   *Prioniodus variabilis* Bergström – Bergström, pp. 147 – 148; Pl. 2: 2.
1976   *Prioniodus variabilis* Bergström – Dzik, Fig. 24h – l.
1977   *Baltoniodus variabilis* (Bergström) – Lindström (*in* Ziegler), pp. 83 – 84, Baltoniodus-plate 2: 13 – 14.
1981   *Prioniodus variabilis* Bergström – Nowlan, p. 12; Pl. 4: 10 – 12, 14 – 17.
1985   *Baltoniodus variabilis* (Bergström) – Bergström & Orchard, p. 58; Pl. 2.3: 2.
1987   *Baltoniodus variabilis* (Bergström) – Bergström, Rhodes & Lindström, Pl. 18.1: 2.
1989   *Baltoniodus variabilis* (Bergström) – Rasmussen & Stouge, Fig. 3D – F.
1994   *Baltoniodus variabilis* (Bergström) – Dzik, p. 84; Pl. 19: 1 – 9, textfigs. 14c, 15.
1994   *Baltoniodus variabilis* (Bergström) – Armstrong, pp. 771 – 772; Pl. 1: 2, Pl. 2: 2 – 9.
1998b *Baltoniodus variabilis* (Bergström) – Zhang, pp. 54 – 55; Pl. 3: 9 – 14.

Remarks. – The individual *Baltoniodus variabilis* elements, except the Pa element, were described by Bergström (1962) and the Pa element by Bergström (1971).

B. variabilis differs from *B. prevariabilis* (Fåhræus) by the presence of the prominent platform ledges on the Pa and Pb elements, and by the distinct outer-lateral flare on Pa elements.

Occurrence. – Høyberget, Glöte. *P. anserinus-A. inaequalis* Subzone.

Material. – 2 Pa, 3 Pb, 7 M, 1 Sa, 5 Sd.

Genus *Coelocerodontus* Ethington, 1959

Type species. – *Coelocerodontus trigonius* Ethington, 1959

Coelocerodontus? aff. *latus* van Wamel, 1974

Not figured

Synonymy. –

aff. 1974 *Coelocerodontus latus* n. sp. – van Wamel, p. 56; Pl. 1: 2a,b.

Remarks. – A single specimen was extracted from the lowermost sample in the Herram section. The specimen is characterised by an angle of 45° between the posterior and anterior margins. The short cusp is distinctly recurved. The element differs from *Coelocerodontus latus* van Wamel by having an anteriorly thickened outer surface. Bagnoli *et al.* (1987) excluded *Coelocerodontus latus* and *C. variabilis* from *Coelocerodontus* and referred them to the new genus *Diaphanodus* instead. Later, Müller & Hinz (1991) regarded *Diaphanodus* as a junior synonym of *Coelocerodontus*.

Occurrence. – Herram. *M. flabellum* - *D. forceps* Zone.

Material. – 1 specimen.

Coelocerodontus? sp.

Pl. 3: 6

Description. – The element is simple, smooth and nongeniculate coniform. The cross-section is planoconvex to asymmetrically biconvex. The anterior margin is rounded, and the upper margin keeled. An anterior carina placed at the convex (outer) side may be present. The walls are thin. The basal cavity extends almost to the tip of the cusp.

Remarks. – *Coelocerodontus?* sp. resembles in some aspects the holotype of *Coelocerodontus? lacrimosus* Kennedy, Barnes & Uyeno 1979 from the *Baltoniodus alobatus* Subzone of New Brunswick, but the former is characterised by a smaller angle between the anterior and the upper margins.

Occurrence. – Andersön-A, Røste. *M. ozarkodella* Zone – *E. suecicus* Zone.

Material. – 4 coniform elements.

Genus *Cornuodus* Fåhræus, 1966

Type species. – *Cornuodus erectus* Fåhræus, 1966

Remarks. – The nongeniculate, coniform multielement genus *Cornuodus* was interpreted as quadrimembrate by Stouge & Bagnoli (1990, p. 14) and comprises symmetrical element A (*Drepanodus longibasis sensu formo*), symmetrical element B, asymmetrical elements and shortbased elements (*Cornuodus bergstroemi sensu formo*). Dzik (1994) interpreted the genus as quinquemembrate.

Cornuodus longibasis (Lindström, 1955)

Pl. 3: 7 – 8

Synonymy. –

1955a *Drepanodus logibasis* n. sp. – Lindström, p. 564; Pl. 3: 31.
1978 *Cornuodus longibasis* (Lindström) – Löfgren, p. 49; Pl. 4: 36, 38 – 42; Text-fig. 25A – C.
1978 *Cornuodus bergstroemi* Serpagli – Löfgren, p. 51; Pl. 4: 37; Text-fig. 25D.
1990 *Cornuodus longibasis* (Lindström) – Stouge & Bagnoli, pp. 13 – 14; Pl. 3: 3 – 7 (*cum. syn.*).
1991 *Cornuodus longibasis* (Lindström) – Rasmussen, pp. 274 – 275.
1994 *Cornuodus longibasis* (Lindström) – Löfgren, Fig. 7: 11, 12.
1994 *Cornuodus longibasis* (Lindström) – Dzik, pp. 61 – 62; Pl. 11: 8 – 13; Text-fig. 4a.
1994 *Cornuodus bergstroemi* Serpagli – Dzik, p. 62; Pl. 11: 14 – 19; Text-fig. 4b.
1995a *Cornuodus longibasis* (Lindström) – Lehnert, pp. 80 – 81.
1995b *Cornuodus longibasis* (Lindström) – Löfgren, Fig. 9an – ar.
1997 *Cornuodus longibasis* (Lindström) – Armstrong, p. 785; Pl. 4: 12 – 22; Text-fig. 6.
1998b *Cornuodus longibasis* (Lindström) – Zhang, pp. 57 – 58; Pl. 1: 15 – 19.

Remarks. – The present elements agree with the descriptions of *C. longibasis* and *C. bergstroemi* by Löfgren (1978).

Occurrence. – Andersön-A, -B, Herram, Steinsodden, Røste, Jøronlia, Haugnes, Hestekinn, Skogstad, Glöte. *P. proteus* Zone – *E. suecicus* Zone.

Material. – 337 specimens.

Genus *Costiconus* n. gen.

Type species. – *Panderodus ethingtoni* Fåhræus, 1966 (costate element)

Derivation of name. – Costatus (Latin), ribbed; conus (Latin), cone. Refer to the outline of the costate elements.

Diagnosis. – The apparatus is basically trimembrate, comprising drepanodontiform, oistodontiform and costate elements. The drepanodontiform and oistodontiform elements are asymmetrical, adenticulate and acostate. The drepanodontiform elements are planoconvex and nongeniculate, while the oistodontiform elements are inwardly bowed, planoconvex or gently biconvex, and geniculate or nongeniculate. The costate elements make a symmetry-transition series of costate, adenticulate, nongeniculate coniform elements making at least 19 different variations based on the location and number of costae on the lateral surfaces. *Costiconus* n. gen. is characterised by a basal cavity which extends about 30 – 50 % of the total length of element.

Discussion. – During the last two decades, a number of reconstructed Ordovician and Silurian conodont apparatuses have been incorporated in *Walliserodus* Serpagli 1967. Fåhræus & Hunter (1985), however, mentioned that the Early and Middle Ordovician "*Walliserodus*" (= *Costiconus* n. gen.) apparatuses might possibly represent another genus. The material presented here does strongly support this, as there are some clear differences between *Walliserodus* and its possible ancestor *Costiconus*.

The costate elements of *Costiconus* have in recent years been referred to the genus *Walliserodus* Serpagli 1967 which has the Silurian *Acodus curvatus* Branson & Branson 1947 as type species (e.g. Bergström *et al.* 1974; Dzik 1976; Löfgren 1978; Nowlan 1981; Stouge 1984; Fåhræus & Hunter 1985a; McCracken 1991; Dzik 1994).

The suggestion that the Lower and Middle Ordovician costate "*Walliserodus*" elements actually belong to the same species as the related drepanodontiform and oistodontiform elements is not new. Löfgren (1978) noted that it was not unlikely that the elements referred by her to *Walliserodus ethingtoni* (Fåhræus) was part of the same apparatus as her new species *Paltodus? jemtlandicus* Löfgren. Nowlan (1981), Stouge (1984), Fåhræus & Hunter (1985a) and Dzik (1991) presented similar ideas but went one step further and included a nongeniculate, acostate element in the "*Walliserodus*" apparatus. Stouge (1984) was of the opinion that the elements referred to *Walliserodus ethingtoni* and *Parapaltodus flexuosus* (Barnes & Poplawski) possibly might be parts of the same apparatus. Zhang (1998b) included P, M and S elements in the apparatus.

Barrick (1977) introduced the following diagnosis of *Walliserodus*: "Elements M, Sa, Sb, Sc, Sd. Elements are thin-walled and basal cavity extends about three-fourth of element height. Costate series comprises strongly costate elements with strongly recurved distal tips. Basal cross-sections vary from symmetrically biconvex (Sd) to symmetrically triangular (Sa) to asymmetrically triangular (Sb) to asymmetrically biconvex (Sc). M element erect with asymmetrical, biconvex outline".

There are, however, clear differences between *Walliserodus* and *Costiconus*:

1. The costate elements of *Costiconus* are associated with two types of acostate elements, herein regarded as drepanodontiform and oistodoniform elements (see descriptions below), whereas *Walliserodus* basically comprises only one acostate element type which is analogous to the drepanodontiform element of *Costiconus*.

2. The basal cavity of *Costiconus* extends only about $\frac{1}{3}$ to $\frac{1}{2}$ of the total length of element, whereas it extends $\frac{2}{3}$ to $\frac{3}{4}$ of the total length of *Walliserodus* elements (cf. Barrick 1977).

3. *Walliserodus* contains about 35% symmetrical elements (Barrick 1977), which is more than *Costiconus*, which in the present material comprises 19% costate elements with equal number of costae on each side of the element, but only 5 % symmetrical elements (Table 1, 2).

Costiconus costatus (Dzik, 1976)

Pl. 3: 9 – 15

Synonymy. –

Drepanodontiform
pt. 1978 *Paltodus?* cf. *mysticus* (Barnes & Poplawski) – Löfgren p. 64; Pl. 4: 11 (only).
    1987 *Paltodus? jemtlandicus* Löfgren – Olgun, Pl. 7: K, T, U.
pt. 1991 *Walliserodus costatus* Dzik – Dzik, Text-fig. 6 (ne element only).
    1997 *Parapaltodus* sp. – Bagnoli & Stouge, Pl. 5: 17 (only).

Oistodontiform
pt. 1978 *Paltodus?* cf. *mysticus* (Barnes & Poplawski) – Löfgren p. 64; Pl. 4: 4, 5 (only).
pt. 1984 *Parapaltodus flexuosus* (Barnes & Poplawski) – Stouge, p. 48; Pl. 1: 22, 24 (only).
    1987 *Paltodus? mysticus* (Barnes & Poplawski) – Olgun, Pl. 7: I, J, L.
    1995b *Walliserodus mysticus* (Barnes & Poplawski) – Löfgren, Fig. 9x (only).

Costate elements
    1976 *Walliserodus costatus* n. sp. – Dzik, p. 421; Pl. XLI: 2; Text-fig. 14m, n.
    1978 *Walliserodus* cf. *ethingtoni* (Fåhræus) – Löfgren, p. 113; Pl. 4: 13, 14.

TABLE I. Number of costae on each of the two lateral surfaces in costate elements of *Costiconus costatus* (Dzik).

Locality	Sample #	Drep.	oist.	costate	0-1	0-2	0-3	1-1 sym	1-1 asym	1-2	1-3	1-4	1-6	2-2 sym	2-2 asym	2-3	2-4	2-5	3-3 sym	3-3 asym	3-4	4-4 asym	4-5	5-6	6-6 asym	indet	drepan. + oist. + cost.
Andersön-A	97812			3																	1						3
	97817			4											2	1				1							4
	97819	1			1																						1
	97825	3	1	3						2						1	1										7
	97827			1											1												1
	97831			2													1										2
Andersön-B	99456	1		1			1			1							1			1							2
	99458	1		2						1	1					1	1										3
	99460	2	2	4							1			2		1	1										8
	99461	9	6	5	1						1				1	1										20
	99464	6				1																					6
	99465			3											2		1										3
	99468	1		1																1							2
	99469	4	1	5						1	1			1	2	2										10
	99470	7	1	7					1	2	2				1	1										15
	99471	11	6	13	2					2					6	2										30
	99472	14	1	17						5	4			2	1	1					1					32
	99473	8	2	25	1			2	1	2	4	3		5	3	6					1					35
	99474	10	2	4						1	1					2										16
	99477		1	3							1			1		1										4
Herram	69685			7						1		1		1		1			1	1	1	1					7
Steinsodden	69687	1	3	4	1											1				1					1	5
	69603	1		1												1										2
	69606																									1
	69608	1		1			1						1													2
	69613	1																								1
	69614	1																								1
	69616	1		4						1					1	1				2						4
	69617	1		1	1																					5
	69621	1																								1
	69622	1	1	3											1											5
	69626			1		1									1	1										1
	69628			2						1					1											2
	69629	16	2	49		1					1			5	18	9	4			1	4		1	1	3	67
	69633	4		16							1			2	2	7					1				2	20
	69635	2	3	4											1	1	2							1		9
	69638	2	2	11											1	7	1			2						15
	69642			3											2	1										3
	69644	5	1	3											2	1										9
	69647	2		11											2	2	3	1		1	2					13
	69649			2								1														2
	69662	5		3												3										8
	69667	3		1												1										5
Røste	97760	4	3	3												1										10
	97763	2	2	5							2					4										9
Haugnes	97674	6	1	8	2					1					1	1	1					1	1			15
	97695		1	1																						2
Hestekinn	97715			1						1	1															1
Jøronlia	97718	1					1																			1
	97719		1	1							1															2
Total		138	44	250	8	4	3	2	2	22	23	5	1	18	52	63	13	1	1	11	11	1	1	1	1	6	432

? 1982 *?Walliserodus ethingtoni* (Fåhræus) *sensu* Löf-
     gren – Ethington & Clark, p. 116; Pl. 13: 10, 14 –
     15 (*non* 16); Text-fig. 35.
1987 *Walliserodus ethingtoni* (Fåhræus) – Olgun, p. 55;
     Pl. 7: C – H.
1991 *Walliserodus costatus* Dzik – Dzik, Text-fig. 6 (all
     but the ne element).
1994 *Walliserodus costatus* Dzik – Dzik, p. 56; Pl. 12:
     1 – 6; Text-fig. 2a.
1995b *Walliserodus mysticus* (Barnes & Poplawski) –
     Löfgren, Fig. 9w (only).
1997 *Parapaltodus* sp. – Bagnoli & Stouge, p.Pl. 5: 16
     (only).

Emended diagnosis. – A *Costiconus* species with an
oistodontiform element that makes an angle of more
than 90° between the upper margin and the posterior
margin of the cusp. The oistodontiform element is gently,
inwardly bowed.

Description. –

Drepanodontiform
The drepanodontiform element is acostate, nongeniculate coniform with a plano-convex to asymmetrical
biconvex base. The element is keeled. The anterior keel is
restricted to the cusp, whereas the posterior keel extends
from the apex of the cusp to the posterior part of the base.
A central, shallow, groove orientated perpendicular to the
basal margin may be present on the convex surface of the
base. The basal cavity excavates the base with a tip that
extends about ⅓ to ½ of the total length of element. The
cusp is proclined.

Oistodontiform
The oistodontiform element is an acostate, planoconvex,
nongeniculate coniform which is flexed inwardly. In
inner lateral view, the angle between the upper margin of
the base and the posterior margin of the cusp is more than
90°. Keels are developed at both the anterior and
posterior margins. Some elements have a central carina
on the inner surface of the cusp. The basal cavity extends
about ⅓ of the total length of element. The cusp is
proclined.

Costate element
The costate elements are costate, nongeniculate coni-
forms which make a symmetry-transition series. The
symmetry-asymmetry is caused by the number and
position of costae. The variety of costate elements is high
(Table 1), varying from elements with 0 costae on one
lateral surface and 1 on the other (0 – 1) to elements with
6 costae on each of the lateral surfaces (6 – 6). The most
common combination is 2 – 3 and asymmetrical 2 – 2 but
the 1 – 2 and 1 – 3 combinations are also fairly common.
The asymmetrical elements are by far the most abundant.

Costate elements have usually one keel on the posterior
margin. The basal cavity is extending ⅓ to ½ of the total
length of element, with the apex situated close to the
anterior margin. The cusp is proclined or erect.

Remarks. – *Costiconus costatus* is distinguished from *C.
ethingtoni* (Fåhræus) by differences in the oistodontiform
element. The angle between the upper margin and the
posterior margin is more than 90° in *C. costatus*. In *C.
ethingtoni* this angle is about 80°. Also, the cusp is
inwardly bowed with a sharp bent between the cusp and
the base in *C. ethingtoni*, whereas the bend is gradual in *C.
costatus*. Löfgren (1978) distinguished two groups of
costate elements: an early (youngest Latorpian and
Volkhovian) form, which she named *Walliserodus* cf.
ethingtoni, characterised by 2 to 3 costae on each side,
longer and less expanded base and a more proclined cusp,
and a later form, the typical *Walliserodus ethingtoni*.
Investigation of the specimens at hand does not support
such a separation, as the costate elements show about the
same variation throughout the studied interval (Latorpian to upper Kundan) (Table 1).

Occurrence. – Andersön-A, -B, Herram, Steinsodden,
Røste, Jøronlia, Haugnes, Hestekinn. *M. flabellum* -
D. forceps Zone – *P. graeai* Zone.

Material. – 138 drepanodontiform, 44 oistodontiform,
250 costate elements.

Costiconus ethingtoni (Fåhræus, 1966)

Pl. 3: 16 – 18

Synonymy. –

Drepanodontiform
1960 *Scandodus* n. sp. 2 – Lindström, Text-fig. 6: 12,
     7: 11.
pt. 1978 *Paltodus? jemtlandicus* n. sp. – Löfgren, p. 65;
     Pl. 4: 6 – 10 (only).
pt. 1984 *Parapaltodus flexuosus* (Barnes & Poplawski) –
     Stouge, p. 48; Pl. 1: 19, 23 (only).

Oistodontiform
pt. 1978 *Paltodus? jemtlandicus* n. sp. – Löfgren, p. 65;
     Pl. 4: 1 – 3 (only).
pt. 1984 *Parapaltodus flexuosus* (Barnes & Poplawski) –
     Stouge, p. 48; Pl. 1: 25 (only).

Costate element
? 1964 *Scolopodus tuatus* n. sp. – Hamar, p. 283; Pl. 2: 5,
     9; Text-fig. 4: 13.
1966 *Panderodus ethingtoni* n. sp. – Fåhræus, p. 26;
     Pl. 3: 5a, b.

TABLE II. Number of costae on each of the two lateral surfaces in costate elements of *Costiconus ethingtoni* (Fåhræus), *Costiconus* cf. *ethingtoni* (Fåhræus) and *Costiconus* sp. A.

Locality	Sample #	drep.	oist.	costate	0-1	0-2	0-3	0-4	0-5	1-1 sym	1-1 asym	1-2	1-3	1-4	2-2 sym	2-2 asym	2-3	2-4	2-5	3-3 sym	3-3 asym	3-4	4-4 asym	Total
C. cf. ethingtoni																								
Andersön-B	99476	17	2	17			1					1			2	4	8		1					36
Steinsodden	69671		1																					1
	69673		1																					1
Røste	97763		1																					1
Jøronlia	97730	2	2	2												2								6
Total		19	7	19	0	0	1	0	0	0	0	1	0	0	2	6	8	0	1	0	0	0	0	45
C. ethingtoni																								
Andersön-B	99478	23	3	13		2	2					2	3				4				1		1	39
	99479	45	7	71						1	5	14	5		8	11	14	4			3	1		123
	99480	2	3	14		1	1	1	1					1		1	5				1			19
	99481	1		2		1								1										3
	99484			2		1										1								2
Andersön-C	99492	5	3	12								2	1		1	4	4							20
Steinsodden	69675	8	6	25		1	1	3	2			4				1	4	1	1		2	4		39
	69676	2		3													2	1						5
Røste	97766	1	1	1									1											3
	97768	14	3	14	1							4			1	3	3			1	1			31
Total		101	26	157	1	6	4	4	3	1	5	26	10	2	10	21	37	6	1	1	8	6	1	284
Costiconus sp. A																								
Andersön-B	99486	12	7	33							1	8	1			14	8				1			52
	99488	2	1	6								2			1		2	1						9
Andersön-C	99496	2	1	10			1					2	1				4				2			13
Total		16	9	49	0	0	1	0	0	0	1	12	2	0	1	14	14	1	0	0	3	0	0	74

1974 *Walliserodus ethingtoni* (Fåhræus) – Bergström, Riva & Kay, Pl. 1: 12.

1976 *Walliserodus ethingtoni* (Fåhræus) – Dzik, Fig. 14o, p.

1978 *Walliserodus ethingtoni* (Fåhræus) – Löfgren, p. 114; Pl. 4: 27 – 35; Text-fig. 33.

1978 *Walliserodus ethingtoni* (Fåhræus) – Tipnis, Chatterton & Ludvigsen; Pl. 9: 23.

? 1981 *Walliserodus ethingtoni* (Fåhræus) – An, Pl. 3: 16.

1984 *Walliserodus ethingtoni* (Fåhræus) – Stouge, p. 64; Pl. 9: 1 – 9.

1985a *Walliserodus ethingtoni* (Fåhræus) – Fåhræus & Hunter, p. 1180; Pl. 3: 11 – 15; Text-fig. 6A – G, ?6 h.

1985 *Walliserodus ethingtoni* (Fåhræus) – An, Du & Gao, Pl. 10: 19 – 22.

1990 *Walliserodus ethingtoni* (Fåhræus) – Pohler & Orchard, Pl. 2: 19.

1991 *Walliserodus ethingtoni* (Fåhræus) – McCracken, p. 52; Pl. 1: 6, 7.

Multielement

1994 *Walliserodus ethingtoni* (Fåhræus) – Dzik, pp. 56 – 58; Pl. 12: 7 – 10, 15 – 19; Text-fig. 2b (the hi element may not belong to this species).

? 1997 *Walliserodus ethingtoni* (Fåhræus) – Armstrong, p. 782; Pl. 5: 10 – 12.

non 1998b *Walliserodus ethingtoni* (Fåhræus) – Zhang, pp. 95 – 96; Pl. 18: 10 – 15.

Description. – The individual *Costiconus ethingtoni* elements were described in detail by Löfgren (1978). The drepanodontiform and oistodontiform elements were described as *Paltodus? jemtlandicus* Löfgren and the costate elements as *Walliserodus ethingtoni* (Fåhræus). The oistodontiform element is geniculate and the angle between the upper margin and the posterior margin of the cusp is about 80°. The cusp is inwardly bowed with a sharp bent between the cusp and the base. As described above, the species is distinguished from *C. costatus* by the sharper angle between the upper margin of the base and the posterior margin of the cusp. Moreover, *C. ethingtoni* is characterised by more well-developed costae on the costate elements (Dzik 1994).

Remarks. – The elements figured by Armstrong (1997) may possibly belong to *Costiconus* sp. A (see this species below) instead of *C. ethingtoni*. The *Costiconus ethingtoni* specimens figured by Zhang (1998b) have very well-developed costae on the costate elements. The oistodontiform element (Zhang 1998b, Pl. 18:10) has an angle between the upper margin of the base and the posterior margin of the cusp of more than 90° and an inwardly bowed cusp with a smooth bend. These characters show that the specimens belong to *Costiconus* sp. A. instead.

Occurrence. – Andersön-B, -C, Steinsodden, Røste. *P. graeai* Zone – *E. suecicus* Zone.

Material. – 106 drepanodontiform, 26 oistodontiform, 157 costate elements.

Costiconus cf. *ethingtoni* (Fåhræus, 1966)

Pl. 4: 1 – 7

Plate IV

1–7: *Costiconus* cf. *ethingtoni* (Fåhræus).

1. Costate element, × 95. Sample 99476, Andersön-B. PMO 165.212.
2. Costate element, × 100. Sample 99476, Andersön-B. PMO 165.213.
3. Costate element, × 80. Sample 99476, Andersön-B. PMO 165.214.
4. Drepanodontiform element, × 35. Sample 99476, Andersön-B. PMO 165.215.
5. Oistodontiform element, × 95. Sample 99476, Andersön-B. PMO 165.216.
6. Oistodontiform element, × 100. Sample 99476, Andersön-B. PMO 165. 217.
7. Drepanodontiform element. Sample 99476, Andersön-B. PMO 165.218.

8–13: *Costiconus* sp. A.

8. Costate element, × 80. Sample 99486, Andersön-B. PMO 165.219.
9. Oistodontiform element, × 65. Sample 99486, Andersön-B. PMO 165.220.
10. Oistodontiform element, × 80. Sample 99486, Andersön-B. PMO 165.221.
11. Costate element, × 100. Sample 99486, Andersön-B. PMO 165.222.
12. Drepanodontiform element, × 65. Sample 99486, Andersön-B. PMO 165.223.
13. Drepanodontiform element, × 75. Sample 99486, Andersön-B. PMO 165.224.

14: *Costiconus* sp. B.

14. Oistodontiform element, × 75. Sample 99491, Andersön-B. PMO 165.225.

15–17: *Dapsilodus? viruensis* (Fåhræus).

15. Acodontiform element, × 100. Sample 99476, Andersön-B. PMO 165.226.
16. Acontiodontiform element, × 70. Sample 99476, Andersön-B. PMO 165.227.
17. Acodontiform element, × 85. Sample 69675, Steinsodden. PMO 165.228.

PLATE IV

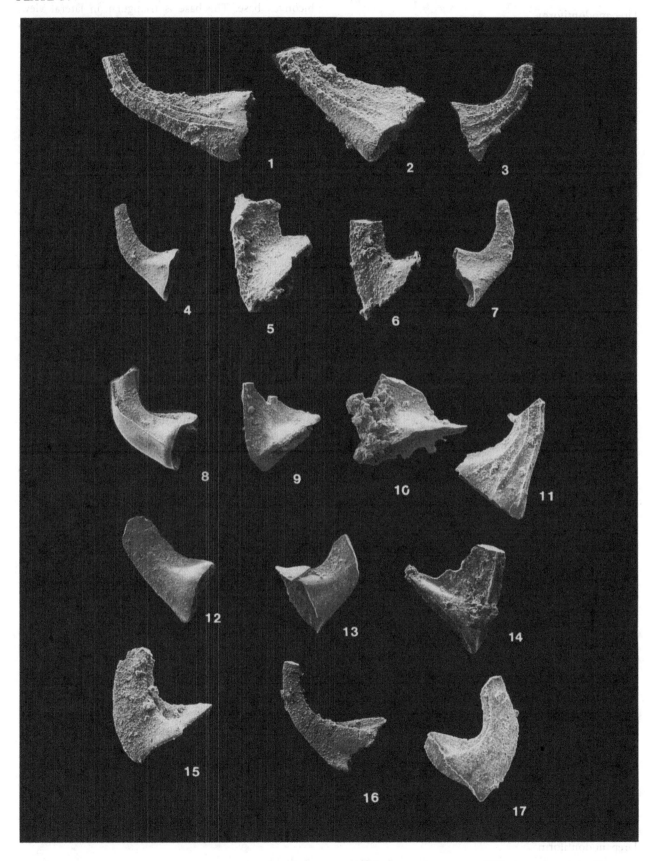

Synonymy. –

Drepanodontiform
? 1973 *Scandodus mysticus* n. sp. – Barnes & Poplawski,
       p. 786; Pl. 4: 1, 2; Text-fig. 2K.

Oistodontiform
? 1973 *Scandodus flexuosus* n. sp. – Barnes & Poplawski,
       p. 785; Pl. 2: 1, 4 (erroneously figured as
       Scandodus inflexus); Text-fig. 2L.

Remarks. – *Costiconus* cf. *ethingtoni* is a transitional form
between *C. costatus* and *C. ethingtoni*, and only the
oistodontiform element separates *C.* cf. *ethingtoni* from
these species. The oistodontiform element is geniculate,
characterised by an angle between the upper margin and
the posterior margin of the cusp that varies from 90 to
115°. The cusp is inwardly bowed with a sharp bent
between the cusp and the base. The species is distinguished
from *C. costatus* by the sharp bent between the base and the
inner surface of the cusp, and from *C. ethingtoni* by the
larger angle between the cusp and the upper margin.

   Scandodus mysticus Barnes & Poplawski and *Scandodus
flexuosus* Barnes & Poplawski possibly are conspecific with
C. cf. *ethingtoni* (Fåhræus), representing the drepanodontiform and oistodontiform element, respectively. If
this is the case, the multielement species should be named
C. flexuosus or *C. mysticus*. The oistodontiform element
(*Scandodus flexuosus sensu formo*; Barnes & Poplawski, p.
785; Pl. 2: 1, 4; Text-fig. 2L) seems to have a smaller basal
cavity than *Costiconus* cf. *ethingtoni*. Accordingly, the
specific assignment has been queried.

Occurrence. – Andersön-B, Steinsodden, Røste, Jøronlia.
M. ozarkodella Zone – *E. suecicus* Zone.

Material. – 19 drepanodontiform, 7 oistodontiform, 19
costate elements.

Costiconus sp. A

Pl. 4: 8 – 13

Synonymy. –

Costate elements
? 1997 *Walliserodus ethingtoni* (Fåhræus) – Armstrong,
       p. 782; Pl. 5: 10 – 12.

Multielement
1998b *Walliserodus ethingtoni* (Fåhræus) – Zhang,
       pp. 95 – 96; Pl. 18: 10 – 15.

Description. –

Drepanodontiform
The drepanodontiform element is acostate, nongeni-

culate coniform with a plano-convex to asymmetrical
biconvex base. The base is triangular in lateral view.
The element is keeled. The anterior keel is restricted to
the cusp. A central, shallow, groove oriented perpendicular to the basal margin may be present on the
convex surface of the base. The cusp is proclined.

Oistodontiform
The oistodontiform element is an acostate, planoconvex
coniform which is flexed inwardly. Keels are developed at
both the anterior and posterior margins. The posterior
keel extends to the tip of the upper margin and is wider
than the anterior keel. In inner lateral view the angle
between the upper margin of the base and the posterior
margin of the cusp is 90° or more. The cusp is inwardly
bowed with a smooth bend. The basal cavity forms an
irregular triangle in an inner-lateral view. The posterior
side of the triangle is clearly longer than the anterior leg.
The cusp is proclined.

Costate elements
The costae are distinct and well-developed. The most
common combination of costae is 2 – 3 and asymmetrical
2 – 2 (19% each) but the 1 – 2 combination are also
common (16%). Symmetrical elements comprises only
3% of the costate elements (Table 2). The basal cavity
extends ⅓ to ½ of the total length of element with the apex
situated close to the anterior margin. The cusp is proclined
or erect.

Remarks. – The oistodontiform of *Costiconus* sp. A differs
from *C. ethingtoni* and *C.* cf. *ethingtoni* by lacking the sharp
bend between the base and the cusp on the inner-lateral
surface. It is distinguished from *C. costatus* by having a
more irregularly triangular base in lateral view, because the
upper-posterior "leg" is clearly longer than the anterior,
and by a more laterally compressed cusp. Generally, the
costate elements have more well-developed costae than the
above-mentioned *Costiconus* species. The three costate
elements referred to *Walliserodus ethingtoni* by Armstrong
(1997) have well-developed, and relatively few costae and
are possibly *Costiconus* sp. A, but it is difficult to give a
more precise species identification when the oistodontiform elements are lacking.

Occurrence. – Andersön-B, Andersön-C. *P. anserinus-S.
kielcensis* Subzone.

Material. – 16 drepanodontiform, 9 oistodontiform and
49 costate elements.

Costiconus sp. B

Pl. 4: 14

Description. – The oistodontiform element is an uni-costate, geniculate coniform which is flexed inwardly. A posterior keel is present. The angle between the upper margin of the base and the cusp is about 80°. The cusp is inwardly bowed with a smooth bend. The basal cavity is shallow. The cusp is proclined.

Remarks. – Only one oistodontiform element was recovered. The species differs from all other *Costiconus* species by the inner-lateral costa, which is placed close to the anterior margin.

Occurrence. – Andersön-B. *N. gracilis* graptolite Zone.

Material. – One oistodontiform element.

Costiconus cf. dolabellus (Fåhræus & Hunter, 1985)

Not figured

Synonymy. –

Costate elements
? 1966 *Panderodus nakholmensis* n. sp. – Hamar, p. 66; Pl. 7: 22 – 24; Text-fig. 3: 3.

Multielement
? 1985a *Walliserodus nakholmensis* (Hamar) – Fåhræus & Hunter, p. 1181; Pl. 2: 1 – 5; Text-fig. 7 (*cum. syn.*).
cf. 1985a *Walliserodus dolabellus* n. sp. – Fåhræus & Hunter, pp. 1179 – 1180; Pl. 2: 6 – 10; Text-fig. 5.
? 1994 *Walliserodus nakholmensis* (Hamar) – Dzik, p. 58; Pl. 12: 11 – 14; Text-fig. 2c.

Comments to the synonymy list. – The costate element differs from the symmetrical multicostate element of "*Walliserodus*" *dolabellus* Fåhræus & Hunter illustrated by Fåhræus & Hunter (1985, Pl. 2: 6a – b; Text-fig. 5a) by the lack of postero-lateral costae but fits in all other aspects with the description of that species. The "inwardly bowed elements" of "*Walliserodus*" *dolabellus* and "*Walliserodus*" *nakholmensis* figured by Fåhræus & Hunter (1985a, p. 2: 5a – b, 9a – b; Text-figs. 5e, 7e) carry an inner-lateral costa, which is lacking in the associated drepanodontiform element at hand. However, the number of specimens is too low to show whether or not this is a matter of intra-specific variation.

Remarks. – One drepanodontiform and one costate element were recovered. The costate element is symmetrical bicostate with two antero-lateral costae and a sharp posterior keel. In all other aspects the costate element conforms with the description of "*W.*" *dolabellus* by Fåhræus & Hunter (1985). It should be

noted that this element type has not been observed in *C. ethingtoni* (Fåhræus).

The drepanodontiform element is similar to that of *Costiconus* sp. A, except that it is slightly more laterally compressed.

Taking the great morphological variety of the costate elements of stratigraphically older *Costiconus* species into consideration (Löfgren 1978, fig. 33) it is possible that *Costiconus nakholmensis* (Hamar) and *C. dolabellus* (Fåhræus & Hunter) represent different elements of the same species.

Occurrence. – Glöte, the *P. anserinus-S. kielcensis* Sub-zone.

Material. – 1 drepanodontiform, 1 costate element.

Genus *Dapsilodus* Cooper, 1976

Type species. – *Distacodus obliquicostatus* Branson & Mehl, 1933

Remarks. – *Dapsilodus* is trimembrate comprising symmetrical acontiodontiform, asymmetrical acontiodontiform, and acodontiform elements (Cooper 1976), which corresponds to the symmetrical p, sq and r elements of Armstrong (1990). According to the emended diagnosis by Armstrong (1990, p. 70), *Dapsilodus* is characterised by a symmetrical p element "with straight upper and lower edges, the former extended laterally by two low costae". *D.? viruensis* does not include such an element, making the generic relationship with *Dapsilodus* dubious. In addition, Silurian and Late Ordovician *Dapsilodus* species differ from the Lower Ordovician *Dapsilodus? viruensis* by the lack of the distinct median, longitudinal striation.

Dapsilodus? viruensis (Fåhræus, 1966)

Pl. 4: 15 – 17

Synonymy. –

Acodontiform
1966 *Acodus viruensis* n. sp. – Fåhræus, p. 12; Pl. 2: 2a – b; Text-fig. 2A.

Acontiodontiform
? 1960 *Acontiodus* n. sp. 1 – Lindström, Text-fig. 5: 15.
1966 *Acontiodus sulcatus* n. sp. – Fåhræus, p. 17; Pl. 2: 6a – b; Text-fig. 2F.

Multielement
1978 *Acodus? mutatus* (Branson & Mehl) – Löfgren, p. 44; Pl. 2: 9 – 21.

1990   *Dapsilodus   mutatus*   (Branson   &   Mehl) –
       Stouge & Bagnoli, p. 14; Pl. 9: 19, 26, 27.

1994   *Dapsilodus viruensis* (Fåhræus) – Dzik, pp. 63 –
       64; Pl. 11: 20 – 23, 27 – 30; Text-fig. 6a – c.

? 1995a *Dapsilodus   mutatus*   (Branson   &   Mehl) –
       Lehnert, p. 81; Pl. 15: 16, Pl. 18: 6.

1998b *Dapsilodus   viruensis*   (Fåhræus) – Zhang,
       pp. 58 – 59; Pl. 4: 1 – 6.

Description. – The acodontiform element is costate on one side, whereas the other side is smooth or carinate with lateral striae placed on the carina. The lateral costa only rarely extends to the basal margin, but if this is the case, the angle between the costa and the basal margin is usually about 50 – 70 degrees. The acontiodontiform elements comprise a symmetry transition series, based on variations in the development of costae. Lateral striations occur on one or both lateral sides. The individual elements were described by Fåhræus (1966) and Löfgren (1978).

Discussion. – Löfgren (1978) referred *Dapsilodus? viruensis* to *Acodus? mutatus* (Branson & Mehl 1933). However, the Scandinavian material is considered here distinct from *Belodus (?) mutatus sensu formo*, based on two basic reasons:

1) Branson & Mehl (1933, p. 126) noted in the original description of the form species *Belodus(?) mutatus* (the acodontiform element) that "The oral edge is set with one or two minute denticles confined to the posterior end (not found in some specimens)". No elements with such denticles were noticed in the Swedish material described by Löfgren (1978), who investigated more than 800 acodontiform elements of "*Acodus? mutatus*" (included in *D.? viruensis* herein), or in the present study.

2) The lateral costa in acodontiform elements of *Dapsilodus? viruensis* usually forms an angle at 50 – 70° with the basal margin. In *Dapsilodus mutatus*, the costa is typically perpendicular to the basal margin.

The antero-basal serrations that characterise the Middle – Late Ordovician *Dapsilodus*-specimens [e.g. *Dapsilodus mutatus? sensu* Orchard 1980, Pl. 5: 6, 16 (only); *D. mutatus sensu* Bergström & Orchard 1985, Pl. 2.5: 17 and *Dapsilodus sensu* Sweet 1988, p. 50: 5.3] occur only very rarely in *D.? viruensis*.

Occurrence. – Andersön-A, -B, -C, Steinsodden, Røste, Jøronlia, Hestekinn, Skogstad, Glöte. *B. medius - H. holodentata* Zone – *P. anserinus-S. kielcensis* Subzone.

Material. – 231 acontiodontiform and 94 acodontiform elements.

Genus *Diaphorodus* Kennedy, 1980

Type species. – *Acodus delicatus* Branson & Mehl, 1933

Plate V

1–3: *Diaphorodus tovei* Stouge & Bagnoli.
1. Sb element, × 130. Sample 97800, Andersön-A. PMO 165.229.
2. Same specimen as Pl. 5:1, but tiltet 45°. × 75.
3. Sa element, × 90. Sample 97800, Andersön-A. PMO 165.230.

4–6: *Drepanodus arcuatus* Pander.
4. Sculponeiform element, × 84. Sample 99479, Andersön-B. PMO 165.231.
5. Arcuatiform element, × 75. Sample 99479, Andersön-B. PMO 165.232.
6. Oistodontiform element, × 55. Sample 69622, Steinsodden. PMO 165.233.

7–8: *Drepanodus planus* Pander.
7. Sculponeiform element, × 70. Sample 99479, Andersön-B. PMO 165.234.
8. Arcuatiform element, × 50. Sample 99479, Andersön-B. PMO 165.235.

9: *Drepanoistodus basiovalis* (Sergeeva).
9. Oistodontiform element, × 125. Sample 69622, Steinsodden. PMO 165.236.

10–16: *Drepanoistodus* cf. *basiovalis* (Sergeeva).
10. Suberectiform element, × 100. Sample 99458, Andersön-B. PMO 165.237.
11. Type 1 element, × 65. Sample 99458, Andersön-B. PMO 165.238.
12. Type 2 element, × 80. Sample 99458, Andersön-B. PMO 165.239.
13. Type 3 element, × 70. Sample 99458, Andersön-B. PMO 165.240.
14. Oistodontiform element, × 100. Sample 99458, Andersön-B. PMO 165.241.
15. Oistodontiform element, × 70. Sample 99458, Andersön-B. PMO 165.242.
16. Oistodontiform element, × 100. Sample 99465, Andersön-B. PMO 165.243.

17: *Drepanoistodus* aff. *basiovalis* (Sergeeva).
17. Oistodontiform element, × 65. Sample 69635, Steinsodden. PMO 165.244.

18–19: "*Drepanoistodus venustus* (Stauffer)".
18. Type C element, × 95. Sample 69647, Steinsodden. PMO 165.245.
19. Oistodontiform element, × 120. Sample 99471, Andersön-B. PMO 165.246.

PLATE V

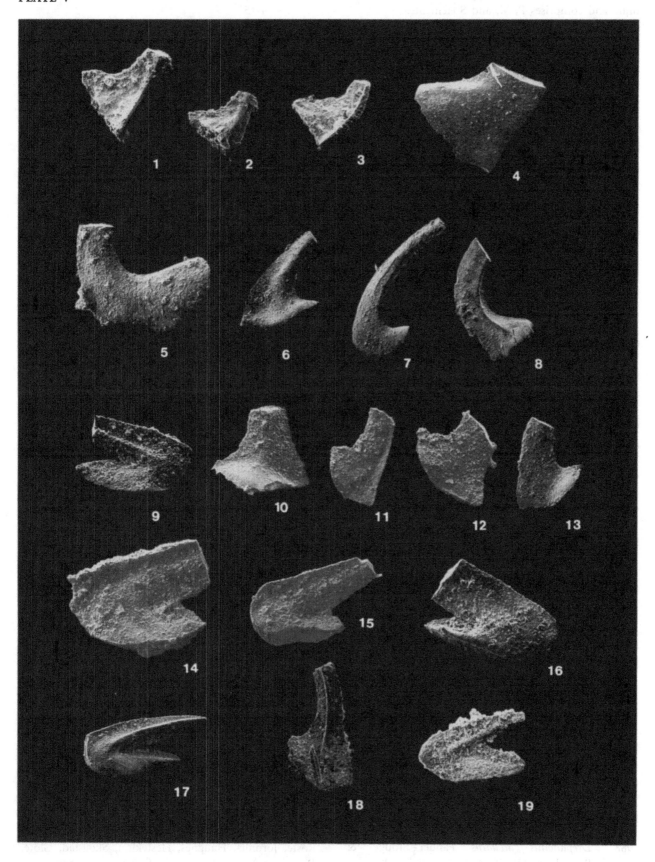

Remarks. – The *Diaphorodus* apparatus is quinquemembrate and comprises P, M, and S elements.

Diaphorodus tovei Stouge & Bagnoli, 1988

Pl. 5: 1 – 3

Synonymy. –

1988 *Diaphorodus tovei* n. sp. – Stouge & Bagnoli, pp. 114 – 115; Pl. 2: 16 – 22 (*cum. syn.*).

Remarks. – *D. tovei* was fully described by Stouge & Bagnoli (1988).

Occurrence. – Andersön-A. *P. proteus* Zone.

Material. – 1 Sa and 1 Sb element.

Genus Drepanodus Pander, 1856

Type species. – *Drepanodus arcuatus* Pander, 1856

Remarks. – The apparatus comprises four nongeniculate elements, that is arcuatiform, sculponeiform, graciliform, and pipaform elements (van Wamel 1974; Stouge & Bagnoli 1988) and an oistodontiform element. Stouge & Bagnoli (1988) separated *Drepanodus arcuatus* and *Drepanodus planus* based on the lack of costae in the former species.

Drepanodus arcuatus Pander, 1856

Pl. 5: 4 – 6

Synonymy. –

Nongeniculate elements
pt. 1856 *Drepanodus arcuatus* n. sp. – Pander, p. 20; Pl. 1: 2, 4, 5, 17 (only).
? 1856 *Drepanodus flexuosus* n. sp. – Pander, p. 20; Pl. 1: 6 – 8, Pl. 3: 4, 11, 12.

Multielement
1974 *Drepanodus arcuatus* Pander – van Wamel, p. 61; Pl. 1: 10 – 13.
pt. 1978 *Drepanodus arcuatus* Pander – Löfgren, p. 51; Pl. 2: 1, 2, 4 – 8 (only).
pt. 1987 *Drepanodus arcuatus* Pander – Olgun, p. 49. Pl. 7A (only).
1988 *Drepanodus arcuatus* Pander – Stouge & Bagnoli, p. 115; Pl. 2: 1 – 6 (*cum. syn.*).

1988 *Drepanodus arcuatus* Pander – Bergström, Pl. 1: 4 – 5.
pt. 1989 *Protopanderodus* n. sp. A – McCracken, p. 23; Pl. 1: 14, 15, 18, 21 (only).
1989 *Drepanodus arcuatus* Pander – Bruton *et al.*, Fig. 5U – W.
1990 *Drepanodus arcuatus* Pander – Stouge & Bagnoli, pp. 14 – 15; Pl. 9: 7 – 10.
1991 *Drepanodus arcuatus* Pander – Rasmussen, p. 275; Fig. 6A – C.
pt. 1994 *Drepanodus arcuatus* Pander – Dzik, pp. 68 – 70; Pl. 15: 2, 3, 6 (only); Text-fig. 8a; 8b (ne, hi, ke only).
cf. 1994 *Drepanodus kielcensis* sp. n. – Dzik, p. 70; Pl. 16: 1 – 7; Text-fig. 9b.
cf. 1994 *Drepanodus santacrucensis* sp. n. – Dzik, pp. 70 – 71; Pl. 16: 8 – 13; Text-fig. 9a.
1994 *Drepanodus arcuatus* Pander – Löfgren, Fig. 6: 35 – 37.
1995a *Drepanodus arcuatus* Pander – Lehnert, p. 82; Pl. 3: 15, 16.
1995b *Drepanodus arcuatus* Pander – Löfgren, Fig. 7: ab – af.
1997 *Drepanodus arcuatus* Pander – Löfgren, Fig. 5: C – I.
1998b *Drepanodus arcuatus* Pander – Zhang, pp. 59 – 60; Pl. 4: 7 – 11, 15, 16.

Description. – The individual element types, except the oistodontiform element, were described by Lindström (1955a).

Oistodontiform element
This element is geniculate coniform with a reclined cusp. The cusp is weakly twisted and is 6 – 8 times longer than the upper margin. The inner side of the cusp is carinate. The upper margin is very short and is weakly inwardly bowed. The basal margin is usually straight. Basal cavity is shallow.

Remarks. – The present material includes an oistodontiform element (Pl. 5: 6) which was not illustrated or described by Stouge & Bagnoli (1988). The element has been found in younger deposits than those studied by Stouge & Bagnoli (*op. cit.*), which makes it possible that the presence of the oistodontiform element *might* be an indicator for the appearance of a new species closely related with *D. arcuatus*. It is also possible, however, that presence or absence of the element was related to environmental differences or geographical differentiation. The geniculate element is possibly a modification of the graciliform element.

Occurrence. – Andersön-A, -B, -C, Herram, Steinsodden, Røste, Jøronlia, Haugnes, Hestekinn, Skogstad, Glöte. *P. proteus* Zone – *P. anserinus-S. kielcensis* Subzone.

Material. – 153 arcuatiform, 12 graciliform, 25 oisto-dontiform, 51 pipaform and 189 sculponeiform elements.

Drepanodus planus (Pander, 1856)

Pl. 5: 7–8

1856　*Machairodus planus* n. sp. – Pander, p. 24; Pl. 2: 39.

pt. 1987　*Drepanodus arcuatus* (Pander) – Olgun, p. 49; Pl. 7B (only).

1988　*Drepanodus planus* (Pander) – Stouge & Bagnoli, p. 116; Pl. 2: 7–10 (*cum. syn.*).

pt. 1989　*Protopanderodus* n. sp. A – McCracken, p. 23; Pl. 1: 11, 12, ?13, 19, 20, 24 (only).

1990　*Drepanodus planus* (Pander) – Stouge & Bagnoli, p. 15; Pl. 9: 11.

1991　*Drepanodus planus* (Pander) – Rasmussen, p. 275; Fig. 6D–E.

pt. 1994　*Drepanodus arcuatus* Pander – Dzik, pp. 68–70; Pl. 15: 4, 5; Text-fig. 8b (pl, tr, oz, sp only).

1998b　*Drepanodus reclinatus* Lindström – Zhang, pp. 60–61; Pl. 4: 12–14, 17–20.

Remarks. – The species was revised and described by Stouge & Bagnoli (1988).

Occurrence. – Andersön-A, -B, Steinsodden, Røste, Jøronlia, Haugnes, Glöte. *P. proteus* Zone – *E. suecicus* Zone.

Material. – 48 arcuatiform, 1 graciliform, 2 oistodonti-form, 10 pipaform and 29 sculponeiform elements.

Genus *Drepanoistodus* Lindström, 1971

Type species. – *Oistodus forceps* Lindström, 1955a

Remarks. – *Drepanoistodus* Lindström is quinquemembrate and comprises 4 nongeniculate and 1 geniculate coniform element types making a curvature-transition series from erect to recurved element types (van Wamel 1974; Cooper 1981; Fåhræus & Hunter 1985b). It is generally difficult to distinguish between the separate *Drepanoistodus* species based on the drepanodontiform and suberectiform elements (van Wamel 1974; Dzik 1976, 1983; Stouge 1984). In the current material this applies to *D. basiovalis* and *D. stougei* whereas *D. forceps* is distinguished by a slightly wider base. "*D. venustus*" and *D. forceps* are the only *Drepanoistodus* species in which costate, nongeniculate elements have been observed (Stouge 1984; Olgun 1987; this study). The nongeniculate elements was separated into suberctiform and drepanodontiform type 1, 2 and 3 (Rasmussen 1991).

Drepanoistodus basiovalis (Sergeeva, 1963)

Pl. 5: 9

Synonymy. –

Oistodontiform

1963　*Oistodus basiovalis* n. sp. – Sergeeva, p. 96; Pl. 7: 6, 7; Text-fig. 3.

Multielement

pt. 1978　*Drepanoistodus basiovalis* (Sergeeva) – Löfgren, pp. 55–56; Pl. 1: ?11–12, 13–16 (only).

pt. 1985　*Drepanoistodus basiovalis* (Sergeeva) – An, Du & Gao, Pl. 13: 17–19.

pt. 1987　*Drepanoistodus basiovalis* (Sergeeva) – Olgun, p. 49; Pl. 6W (only).

1990　*Drepanoistodus basiovalis* (Sergeeva) – Stouge & Bagnoli, p. 15; Pl. 5: 18–24 (*cum. syn.*).

1994　*Drepanoistodus basiovalis* (Sergeeva) – Dzik, p. 78; Pl. 16: 16–20; Text-fig. 12a.

1994　*Drepanoistodus basiovalis* (Sergeeva) – Löfgren, Fig. 6: 30.

non 1995a　*Drepanoistodus basiovalis* (Sergeeva) – Lehnert, pp. 83–84; Pl. 9: 9, Pl. 10: 5, 12.

non 1995b　*Drepanoistodus basiovalis* (Sergeeva) – Löfgren, Fig. 9: y–ad.

1998b　*Drepanoistodus basiovalis* (Sergeeva) – Zhang, pp. 61–62; Pl. 5: 5–12.

Remarks. – Stouge & Bagnoli (1990) distinguished between *D. basiovalis* (Sergeeva) *sensu stricto* and *D. cf. basiovalis*, by the fact that the oistodontiform element of the former species has a relatively long upper margin ("almost ⅓ of the cusp length" *sensu* Stouge & Bagnoli 1990). In contrast, *D. cf. basiovalis* has a shorter upper margin. It may be added that *D. basiovalis* usually has a distinct, inner carina on the cusp, whereas *D. cf. basiovalis* often is smooth or has a weakly developed inner carina. A restudy of the conodonts from the Huk Formation (Rasmussen 1991) confirms that *D. basiovalis* becomes more common than *D. cf. basiovalis* in the *M. parva* Zone and higher, as indicated by Stouge & Bagnoli (1990). See Stouge & Bagnoli (1990) for description. The specimens figured by Lehnert (1995a, Pl. 9: 9, Pl. 10: 5, 12) are typified by a rounded upper anterior edge, which is not characteristic of *D. basiovalis*. They possibly belong to *Paroistodus*. The oistodontiform elements shown by Löfgren (1995b, Fig. 9: ab, ad) have

PLATE VI

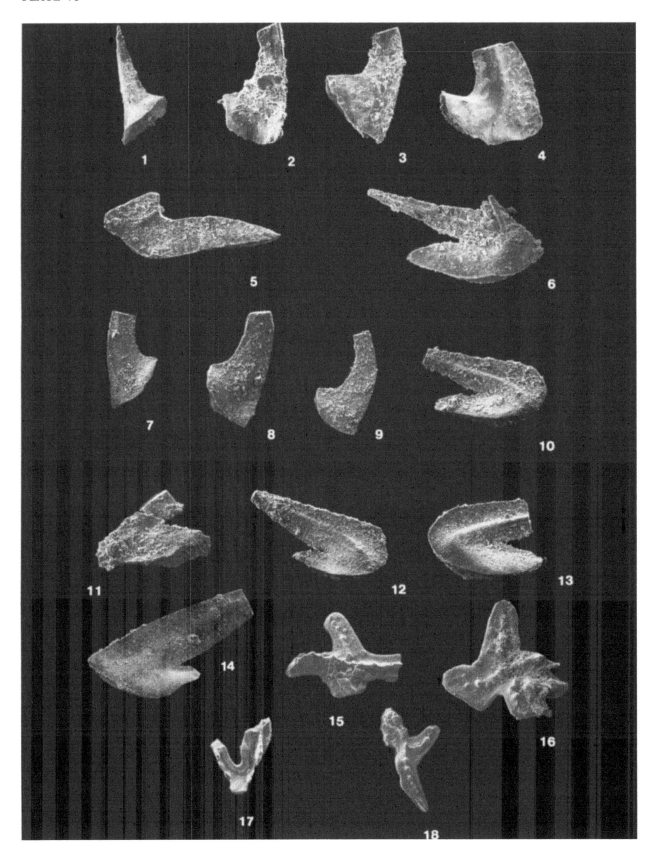

a relatively short upper margin, which is typical of *D. cf. basiovalis*. The figured elements are from her sample Öl 94-9, which is from a level just below the first occurrence of *Microzarkodina parva*. This fits well with the data presented here, where *D. basiovalis s.s.* appears at, or a little above, the first occurrence of *M. parva*.

Occurrence. – Andersön-A, -B, Steinsodden, Røste. *P. rectus - M. parva Zone – P. graeai* Zone.

Material. – 11 suberectiform, 3 drepanodontiform type 1, 12 drepanodontiform type 2, 56 drepanodontiform type 3 and 36 oistodontiform elements.

Drepanoistodus cf. *basiovalis* (Sergeeva, 1963)

Pl. 5: 10 – 16

Synonymy. –

Drepanodontiform type 1

Plate VI

1–4, 6: *Drepanoistodus forceps* (Lindström).
1. Suberectiform element, × 65. Sample 69687, Steinsodden. PMO 165.247.
2. Type 1 element, × 140. Sample 69687, Steinsodden. PMO 165.248.
3. Type 2 element, × 95. Sample 69687, Steinsodden. PMO 165.249.
4. Type 3 element, × 95. Sample 69687, Steinsodden. PMO 165.250.
6. Oistodontiform element, × 145. Sample 69687, Steinsodden. PMO 165.251.

5: "*Drepanoistodus venustus* (Stauffer)".
5. Oistodontiform element, × 170. Sample 99471, Andersön-B. PMO 165.252.

7–10, 13: *Drepanoistodus stougei* Rasmussen.
7. Type 1 element, × 70. Sample 69628, Steinsodden. PMO 165.253.
8. Type 2 element, × 85. Sample 69628, Steinsodden. PMO 165.254.
9. Type 3 element, × 60. Sample 69628, Steinsodden. PMO 165.255.
10. Oistodontiform element, × 85. Sample 69628, Steinsodden. PMO 165.256.
13. Oistodontiform element, × 100. Sample 69628, Steinsodden. PMO 165.257.

11: "*Drepanoistodus* cf. *venustus* (Stauffer)".
11. Oistodontiform element, × 120. Sample 99486, Andersön-B. PMO 165.258.

12: *Drepanoistodus* cf. *stougei* Rasmussen.
12. Oistodontiform element, × 70. Sample 69616, Steinsodden. PMO 165.259.

14: *Drepanoistodus tablepointensis* Stouge.
14. Oistodontiform element, × 85. Sample 69635, Steinsodden. PMO 165.260.

15–18: *Eoplacognathus lindstroemi* (Hamar).
15. Pa element, × 70. Sample 99486. Andersön-B. PMO 165.261.
16. Pa element, × 65. Sample 99486. Andersön-B. PMO 165.262.
17. Pb element, ?dextral, × 70. Sample 99486, Andersön-B. PMO 165.263.
18. Pb element, dextral, × 70. Sample 99486, Andersön-B. PMO 165.264.

1967 *Drepanodus homocurvatus* Lindström – Viira, Text-fig. 1: 16.

Drepanodontiform type 3
1967 *Drepanodus planus* Lindström – Viira, Text-fig. 1: 17.

Multielement
1971 *Drepanoistodus basiovalis* (Sergeeva) – Lindström, p. 43; Figs. 6, 8.
pt. 1978 *Drepanoistodus basiovalis* (Sergeeva) – Löfgren, p. 55; Pl. 1: 17 (only).
pt. 1980 *Drepanoistodus forceps* (Lindström) – Merrill, Fig. 6: 20 – 23 (only).
pt. 1983 *Drepanoistodus forceps* (Lindström)? – Dzik, p. 337; Text-fig. 8b (only the middle element); non Pl. 3: 1 – 4.
1984 *Drepanoistodus basiovalis* (Sergeeva) – Stouge, p. 53; Pl. 3: 18 – 20.
pt. 1987 *Drepanoistodus basiovalis* (Sergeeva) – Olgun, p. 49; Pl. 6: X – EF; non W (= D. basiovalis), GH (= Paroistodus originalis).
1990 *Drepanoistodus* cf. *basiovalis* (Sergeeva) – Stouge & Bagnoli, pp. 15 – 16; Pl. 5: 10 – 16 (cum. syn.).
non 1991 *Drepanoistodus* cf. *basiovalis* (Sergeeva) – McCracken, pp. 49 – 50; Pl. 3: 10 – 12, 15 – 18.
pt. 1991 *Drepanoistodus basiovalis* (Sergeeva) – Rasmussen, p. 277; Fig. 6L (only).
1995b *Drepanoistodus* cf. *basiovalis* (Sergeeva) – Löfgren, Fig. 7z – aa.
1995b *Drepanoistodus basiovalis* (Sergeeva) – Löfgren, Fig. 9y – ad.

Remarks. – *Drepanoistodus* cf. *basiovalis* was described by Stouge & Bagnoli (1990). See *D. basiovalis* for additional remarks. *Drepanoistodus* cf. *basiovalis sensu* McCracken (1991) is different from *D.* cf. *basiovalis sensu* Stouge & Bagnoli (1990) because the anterior margin of the oistodontiform element points slightly downward instead of slightly upward. Also, it usually has a longer upper margin.

Occurrence. – Andersön-B and Steinsodden. *M. flabellum - D. forceps* Zone – *B. medius - H. holodentata* Zone.

Material. – 8 suberectiform, 6 drepanodontiform type 1, 15 drepanodontiform type 2, 29 drepanodontiform type 3 and 27 oistodontiform elements.

Drepanoistodus aff. *basiovalis* (Sergeeva, 1963)

Pl. 5: 17

Description. – The drepanodontiform type 3 element agrees with the description of *Drepanoistodus stougei*.

Oistodontiform
The element is geniculate coniform with a reclined cusp. The lower margin of the cusp is 1.3–1.5 times longer than the upper margin of the base. A well-developed carina is placed on the inner side of the cusp. The upper margin is long with a distinctive flare. The basal margin is slightly convex. The basal cavity is shallow and commonly inverted.

Remarks. – Oistodontiform elements of *Drepanoistodus* aff. *basiovalis* differ from those of *D. basiovalis* by the longer upper margin and a sharper carina, from those of *D. stougei* by the longer upper margin and a more straight anterior margin (instead of a rounded), and from those of "*D. venustus*" by the inverted basal cavity and a more straight anterior margin.

Occurrence. – Steinsodden, Haugnes. *B. norrlandicus - D. stougei* Zone – *B. medius - H. holodentata* Zone.

Material. – 1 drepanodontiform type 3 and 7 oistodontiform elements.

Drepanoistodus forceps (Lindström, 1955)

Pl. 6: 1–6

Synonymy. –

Suberectiform
1955a *Drepanodus suberectus* (Branson & Mehl) – Lindström, p. 568; Pl. 2: 21–22.

Drepanodontiform, type 1
1955a *Acodus gratus* n. sp. – Lindström, p. 545; Pl. 2: 27–29.

Drepanodontiform, type 2
1955a *Drepanodus homocurvatus* n. nom. – Lindström, p. 563; Pl. 2: 23, 24, 39; textfigs. 4d.

Drepanodontiform, type 3
1955a *Drepanodus planus* n. sp. – Lindström, p. 565; Pl. 2: 35–37; Text-fig. 4a.

Oistodontiform
1955a *Oistodus forceps* n. sp. – Lindström, p. 574; Pl. 4: 9–13; Text-fig. 3m.

Multielement
1971   *Drepanoistodus forceps* (Lindström) – Lindström, p. 42; Figs. 5, 8.
1978   *Drepanoistodus forceps* (Lindström) – Löfgren, p. 53, Pl. 1: 1–6; Fig. 26A.

non 1978   *Drepanoistodus forceps* (Lindström) – Fåhræus & Nowlan, p. 459; Pl. 1: 22–25.
non 1981   *Drepanoistodus forceps* (Lindström) – An, Pl. 3: 7, 12, 13.
pt. 1983   *Drepanoistodus forceps* (Lindström) ? – Dzik, p. 337; Text-fig. 8a, *non* Pl. 3: 1–4, *non* Text-fig. 8b.
non 1984   *Drepanoistodus forceps* (Lindström) – Stouge, p. 53; Pl. 3: 24–25.
non 1985   *Drepanoistodus forceps* (Lindström) – Löfgren, Figs. 4AC–AE.
1985   *Drepanoistodus forceps* (Lindström) – An, Du & Gao, Pl. 8: 3, 7, 17.
1987   *Drepanoistodus forceps* (Lindström) – Olgun, Pl. 6: Q–V.
pt. 1988   *Drepanoistodus forceps* (Lindström) – Bergström, Pl. 1: 6–7, 9–10 (only).
1990   *Drepanoistodus forceps* (Lindström) – Stouge & Bagnoli, pp. 16–17; Pl. 5: 6–9 (*cum. syn.*).
1991   *Drepanoistodus forceps* (Lindström) – Rasmussen, p. 277; Fig. K.
1994   *Drepanoistodus forceps* (Lindström) – Löfgren, Fig. 6: 19–22.
pt. 1995a *Drepanoistodus forceps* (Lindström) – Lehnert, p. 84; Pl. 3: 20, Pl. 6: 2, *non* Pl. 13: 2.
1995b   *Drepanoistodus forceps* (Lindström) – Löfgren, Fig. 7ag–al.

Comments to the synonymy list. – The geniculate element illustrated An (1981) is characterised by a sharp costa, and it does not belong to *D. forceps*. Van Wamel (1974) and Dzik (1976, 1983) did not distinguish between *D. forceps* and *D. basiovalis*. Dzik (1983, p. 337) included both *D. forceps*, *D. basiovalis* and *D.*? cf. *venustus sensu* Löfgren (= *D. stougei*) in one species that he named *Drepanoistodus forceps* (Lindström)? He concluded that the morphological differences between the taxa were too small to separate into more than one species without detailed biometrical analysis. In the present material, it has been possible to separate *D. forceps*, *D. basiovalis*, *D.* cf. *basiovalis* and *D. stougei* on the oistodontiform element.

The elements figured by Stouge (1984, Pl. 3: 24, 25) do not belong to *D. forceps* as they lack the distinct carina on the cusp.

The oistodontiform element of Löfgren (1985, Fig. 4AC) has a short upper margin and a more straight anterior than typical found in *D. forceps*. The geniculate element included in *D. forceps* by Bergström (1988, Pl. 1: 8) has a very short upper margin, and is likely part of *Drepanoistodus contractus* (Lindström) *sensu* Stouge & Bagnoli (1990). The element figured by Lehnert (1995a, Pl. 13: 2) is probably the oistodontiform element of *Paroistodus originalis*.

Remarks. – *Drepanoistodus forceps* was fully described by Lindström (1955a, 1971, in Ziegler 1977). The geniculate element of *D. forceps* may be distinguished from "*Drepanoistodus venustus* (Stauffer)" by an undulating basal margin of the geniculate element (Lindström 1971) and a more rounded inner costa on the cusp (Löfgren 1978).

Lindström (1955a, p. 576) pointed out that the *D. forceps* cusp is almost twice as long as the base, whereas the cusp and base of *D. venustus* are equally long. In the present work it appears that the base and cusp are equally long in the upper Llanvirn to Caradoc "*D. venustus*", whereas the cusp is clearly longer than the base in the lower Llanvirn specimens of "*D. venustus*".

Occurrence. – Andersön-A, -B, Herram, Steinsodden. *P. proteus* Zone – *M. flabellum* - *D. forceps* Zone.

Material. – 34 suberectiform, 21 drepanodontiform type 1, 41 drepanodontiform type 2, 244 drepanodontiform type 3 and 293 oistodontiform elements.

Drepanoistodus stougei Rasmussen, 1991

Pl. 6: 7 – 10, 13

Synonymy. –

Multielement
pt. 1985　*Drepanoistodus basiovalis* (Sergeeva) – An, Du & Gao, Pl. 13: 18?, 20 (only).
1991　*Drepanoistodus stougei* n. sp. – Rasmussen, pp. 277 – 278; Fig. 6F – J, N (*cum. syn.*).

Description. – See Rasmussen (1991) for the original description of *Drepanoistodus stougei*. Collected material in the present work is more abundant and better preserved than that published from Slemmestad (Rasmussen 1991), allowing for a more precise description.

Suberectiform
This element is nongeniculate coniform with an erect cusp. The upper margin is short and slightly concave, while the basal margin is convex. The cusp is laterally compressed having sharp anterior and posterior edges. An indistinct carina may occur laterally on the cusp. The element is subsymmetrical around the anterior-posterior vertical plane. The basal cavity is shallow with a posteriorly directed apex.

Drepanodontiform, type 1
This element is nongeniculate coniform with a strongly recurved cusp. The cusp has sharp posterior and anterior edges. The anterior edge – twisted strongly inwards – continues commonly in a lateral, downward-directed costa over the proximal part of the element. Some specimens have a more medial costa. The upper margin is short with a narrow, nearly straight keel. The basal margin is convex or almost straight. The basal cavity is almost cone-shaped.

Drepanodontiform, type 2
This element is nongeniculate coniform with a recurved cusp. The cusp has commonly sharp anterior and posterior edges. A triangular flare is located in the antero-basal corner. The anterior margin may be inwardly flexed, but the sharp bent which typifies the type 1 element is missing. The upper margin is sharp and weakly concave or straight. The basal margin forms a smooth, convex curve. The basal cavity is relatively deep and is almost triangular in a lateral view.

Drepanodontiform, type 3
The type 3 element is nongeniculate coniform with a weakly recurved cusp. The cusp is laterally compressed with sharp anterior and posterior edges. The upper margin is sharp and straight. The basal margin is strongly concave in the posterior part, but is straight and parallel with the upper margin anteriorly. The basal cavity is shallow with an anteriorly directed apex.

Oistodontiform
This element is geniculate coniform with a reclined cusp. The cusp is weakly twisted and is 1.5 to 2 times longer than the base. A distinct carina is located on the inner side of the cusp. The upper margin is long with a distinctive flare. The basal margin is slightly convex. The basal cavity is shallow and may be weakly inverted.

Remarks. – *D. stougei* is distinguished from *Drepanoistodus* cf. *basiovalis* by the longer upper margin (about 50 to 80% of the cusp length) and the rounded anterior margin on the oistodontiform element. Beside the localities mentioned in the synonymy, *D. stougei* has also been reported from the Tartu-453 drillcore of the South Estonian Confacies Belt, Estonia (Stouge *in* Põldvere *et al.* 1998; appendix 13).

Occurrence. – Andersön-A, -B, Steinsodden, Haugnes. *B. norrlandicus* - *D. stougei* Zone and the basal part of the *B. medius* - *H. holodentata* Zone.

Material. – 31 suberectiform, 58 drepanodontiform type 1, 64 drepanodontiform type 2, 185 drepanodontiform type 3 and 161 oistodontiform elements.

Drepanoistodus cf. *stougei* Rasmussen, 1991

Pl. 6: 12

Synonymy. –

1980 *Drepanoistodus forceps* (Lindström) – Merrill, Fig. 20 – 25.

Remarks. – The description of *Drepanoistodus* cf. *stougei* conforms in every aspect with that of *D. stougei*, except that the upper margin of the base in the oistodontiform element is relatively shorter and only constitutes about ⅓ of the length of the lower margin of the cusp.

Occurrence. – Steinsodden. *P. rectus - M. parva* Zone.

Material. – 2 suberectiform, 1 drepanodontiform type 1, 2 drepanodontiform type 2, 6 drepanodontiform type 3 and 1 oistodontiform element.

Drepanoistodus tablepointensis Stouge, 1984

Pl. 6: 14

Synonymy. –

Multielement

1984 *Drepanoistodus tablepointensis* n. sp. – Stouge, p. 54; Pl. 4: 9 – 17 (*cum. syn.*).
1991 *Drepanoistodus tablepointensis* Stouge – Rasmussen, p. 278; Fig. 6O – P.

Remarks. – *Drepanoistodus tablepointensis* was fully described by Stouge (1984). The species previously was reported from North America (Stouge 1984) and the Oslo Region, Norway (Rasmussen 1991).

Occurrence. – Andersön-B, Steinsodden. *B. medius - H. holodentata* Zone – *P. graeai* Zone.

Material. – 1 suberectiform, 3 drepanodontiform type 3, 3 oistodontiform elements.

"Drepanoistodus venustus (Stauffer, 1935)"

Pl. 5: 18 – 19, Pl. 6: 5

Synonymy. –

Oistodontiform
1935 *Oistodus venustus* n. sp. – Stauffer, p. 146. Pl. 12: 12.

Multielement
1978 *Drepanoistodus? venustus* (Stauffer) – Löfgren, p. 57; Pl. 1: 9 – 10 (*cum. syn.*).
1981 *Drepanoistodus? cf. venustus* (Stauffer) – Nowlan, p. 11; Pl. 1: 13, Pl. 3: 7, 17.
1984 *Drepanoistodus? cf. venustus* (Stauffer) – Stouge, p. 55; Pl. 4: 18 – 25.
1983 *Paroistodus? mutatus* (Branson & Mehl) – Nowlan, Pl. 2: 20.
1985 *Drepanoistodus? cf. venustus* (Stauffer) – An, Du & Gao, Pl. 13: 15.
1987 *Drepanoistodus? venustus* (Stauffer) – Olgun, pp. 49 – 50.
1989 *"Oistodus" venustus* Stauffer – Rasmussen & Stouge, Fig. 3L, O.
1990 *Drepanoistodus venustus* (Stauffer) – Stouge & Bagnoli, p. 17; Pl. 5: 17.
1990 *"Oistodus" cf. venustus* Stauffer – Bergström, Pl. 2: 10, Pl. 3: 13.
1991 *Drepanoistodus venustus* (Stauffer) – Rasmussen, p. 278; Fig. 6M.

Plate VII

1–5: *Eoplacognathus reclinatus* (Fåhræus).
1. Pa element, × 40. Sample 99484, Andersön-B. PMO 165.265.
2. Pa element, × 35. Sample 99484, Andersön-B. PMO 165.266.
3. Pb element, de × tral, × 50. Sample 99484, Andersön-B. PMO 165.267.
4. Pb element, de × tral, × 70. Sample 99484, Andersön-B. PMO 165.268.
5. Pb element, sinistral, × 60. Sample 99484, Andersön-B. PMO 165.269.

6–9, 13: *Eoplacognathus suecicus* Bergström.
6. Pa element, × 60. Sample 99479, Andersön-B. PMO 165.270.
7. Pb element, sinistral, × 100. Sample 99479, Andersön-B. PMO 165.271.
8. Pb element, de × tral, × 60. Sample 99479, Andersön-B. PMO 165.272.
9. Pb element, de × tral, × 70. Lateral view. Sample 69675, Steinsodden. PMO 165.273.
13. Same specimen as Pl. 7:9. Upper view. × 75.

10–11, 14: *Erraticodon* sp.
10. Pa element, × 100. Sample 99486, Andersön-B. PMO 165.274.
11. Pa element, × 130. Sample 99486, Andersön-B. PMO 165.275.
14. S element, × 130. Sample 99486, Andersön-B. PMO 165.276.

12: *Polonodus?* sp. B.
12. P element, × 50. Sample 69667, Steinsodden. PMO 165.277.

15: *Fahraeusodus* aff. *marathonensis* (Bradshaw).
15. P element, × 85. Sample 99453, Andersön-B. PMO 165.278.

16: *Fahraeusodus marathonensis* (Bradshaw).
16. S element, × 185. Sample 69635, Steinsodden. PMO 165.279.

17: *Goverdina?* sp.
17. ?S element, × 110. Sample 99473, Andersön-B. PMO 165.280.

18–19: *Histiodella holodentata* Ethington & Clark.
18. P element, × 145. Sample 69668, Steinsodden. PMO 165.281.
19. P element, × 150. Sample 69668, Steinsodden. PMO 165.282.

PLATE VII

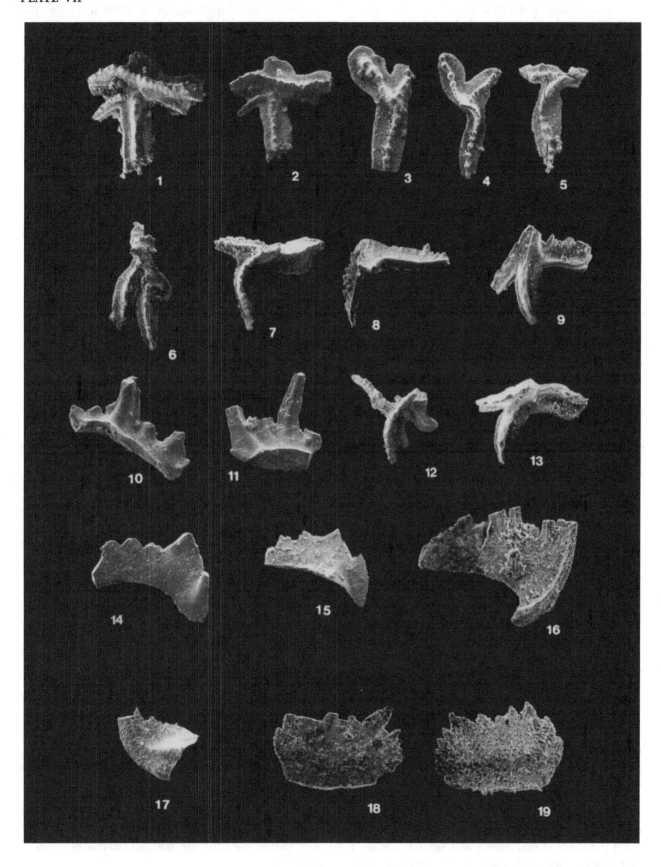

1991 *Drepanoistodus* cf. *venustus* (Stauffer) –
McCracken, p. 46; Pl. 3: 7.

1994 *Paltodus? venustus* (Stauffer) – Dzik, pp. 76 – 78;
Text-fig. 12d.

1995a *Drepanoistodus venustus* (Stauffer) – Lehnert,
p. 84; Pl. 15: 13.

1998b *Drepanoistodus? venustus* (Stauffer) – Zhang,
p. 62; Pl. 5: 13 – 14.

Description. –

Drepanodontiform and suberectiform elements
It has not been possible to separate the acostate
drepanodontiform and suberectiform elements from
other *Drepanoistodus* elements (see also Stouge &
Bagnoli 1990, p. 17). Costate elements seem to be
restricted to "*D. venustus*" and *D. forceps*. Costate
drepanodontiform type 1 and 3 elements occur
sporadically, whereas no costate suberectiform or
drepanodontiform type 2 elements were observed.
Type 1 and type 3 elements are unicostate or multicostate. The base is commonly slightly wider than the
base in acostate elements.

Oistodontiform
The oistodontiform element is geniculate coniform
with a long upper margin. Distinct keels occur at the
upper margin and at the lower edge of the cusp. The
basal margin is rounded. The anterior margin forms
an acute angle that is directed slightly upward. A
distinct, median carina is developed on both sides of
the cusp.

Remarks. – Lindström (1971) included *Oistodus venustus sensu formo* in the *Drepanoistodus* apparatus, while
Barnes and Poplawski (1973) suggested that it
comprises the oistodontiform element in the *Dapsilodus mutatus* apparatus. The present material strongly
supports the reconstruction of Lindström (1971), as
"*O. venustus s.f.*" appears 8 metre below *D. mutatus* in
the Stein section. Löfgren (1978) and Stouge & Bagnoli
(1990) reached the same conclusion. Costate drepanodontiform *D. venustus* elements have not been
recorded from the Baltoscandian inner shelf deposits
(Löfgren 1978; Stouge & Bagnoli 1990; Rasmussen
1991), whereas it was documented from ocean-near
environments at both sides of the Iapetus Ocean
(Stouge 1984 and this work). Consequently, it is
possible that "*D. venustus*" may be separated into two
(?eco-) species, based on the occurrence of costate,
drepanodontiform elements. It is doubtful if the
Scandinavian occurrences of *D. venustus* really
belong to this species, which originally was described
from the United States. Bergström (written comm.
1997) presumed it may belong to a separate species. I
acknowledge this, but because the two forms are very

similar, and that the material at hand does not permit
a test of the hypothesis (it is difficult to see the
internal structures because of thermal alteration), the
species name *venustus* has been put in inverted
comma's, to indicate the ambiguous species
classification.

Occurrence. – Andersön-A, -B, -C, Steinsodden, Røste,
Jøronlia, Haugnes, Hestekinn, Glöte, Engerdal,
Høyberget, Sorken. Upper part of *B. norrlandicus -
D. stougei* Zone – *P. anserinus-A. inaequalis* Subzone.

Material. – 34 suberectiform, 42 drepanodontiform type
1, 33 drepanodontiform type 2, 229 drepanodontiform
type 3 and 279 oistodontiform elements (includes 1
costate suberect., 3 costate drep. type 1 and 1 costate
drep. type 3).

"*Drepanoistodus* cf. *venustus* (Stauffer, 1935)"

Pl. 6: 11

Remarks. – The nongeniculate elements differ from the
above-mentioned *Drepanoistodus* spp. by having wider
bases but otherwise, the description of the nongeniculate elements agrees with these (see description of
Drepanoistodus stougei). The base-cusp transition forms
a sharp bent in the suberectiform element.

The oistodontiform elements are distinguished from
"*Drepanoistodus venustus*" by a wider base and a
relatively shorter upper margin. The cusp is slender
and the inner carina is indistinct and rounded.

Occurrence. – Andersön-B. *P. anserinus-S. kielcensis*
Subzone.

Material. – 1 suberectiform, 1 drepanodontiform type 1,
2 drepanodontiform type 2, 10 drepanodontiform type
3 and 5 oistodontiform elements.

Genus *Eoplacognathus* Hamar, 1966

Type species. – Ambalodus lindstroemi* Hamar, 1964

Remarks. – The form-genus *Eoplacognathus* comprises
the Y- or T-shaped elements, which are interpreted to
represent the ambalodontiform (Pb) elements in the
multielement apparatus (Bergström 1971). Bergström
(1971, in Robison 1981) included four morphotypes in
the apparatus: Sinistral and dextral Pa elements
and sinistral and dextral Pb elements. The sinistral
and dextral elements are not mirror images of each
other.

Eoplacognathus lindstroemi (Hamar, 1964)

Pl. 6: 15–18

Synonymy. –

Pa

1964 *Polyplacognathus elongata* (Bergström) – Hamar, pp. 275–276; Pl. 6: 3, 6–10; Text-fig. 5: 7a–b, 9a–c.

1960 *Amorphognathus* n. sp. 3 – Lindström, Text-fig. 7: 13–14.

1974 *Polyplacognathus mirus* n. sp. – Viira, pp. 107–108; Pl. 9: 10–12, 21; Text-fig. 136.

1974 *Polyplacognathus gallus* n. sp. – Viira, p. 106; Pl. 9: 19–20, 27–28, 31–32; Text-fig. 134.

1974 *Polyplacognathus stella* n. sp. – Viira, pp. 111–112; Pl. 9: 22–26, 29–30; Text-fig. 142–143.

Pb

1964 *Ambalodus lindstroemi* n. sp. – Hamar, pp. 258–259; Pl. 5: 1, 4, 7–8, 10–11; Text-fig. 5: 1a–b, 3a–b, 4a–b.

1964 *Ambalodus* aff. *lindstroemi* n. sp. – Hamar, pp. 259–260; Pl. 5: 5, 12; Text-fig. 5: 2a–b.

1967 *Eoplacognathus lindstroemi* (Hamar) – Viira, Text-fig. 4: 5a–b.

1974 *Eoplacognathus lindstroemi* (Hamar) – Viira, pp. 74–75; Pl. 8: 14, ?15, 19–22; Text-fig. 81.

Multielement

1971 *Eoplacognathus lindstroemi* (Hamar) – Bergström, p. 139; Pl. 2: 15–18.

1974 *Eoplacognathus lindstroemi* (Hamar) – Bergström, Riva & Kay, Pl. 1: 14–15.

1976 *Eoplacognathus lindstroemi* (Hamar) – Dzik, p. 433; Fig. 33a–c.

1985 *Eoplacognathus lindstroemi* (Hamar) – Bergström & Orchard, p. 58; Pl. 2.2: 11, 13.

1994 *Eoplacognathus lindstroemi* (Hamar) – Dzik, p. 98; Pl. 21: 6–9; Text-fig. 23.

1995a *Eoplacognathus lindstroemi* (Hamar) – Lehnert, pp. 86–87; Pl. 15: 19, Pl. 16: 7.

1997 *Eoplacognathus lindstroemi* (Hamar) – Armstrong, p. 772; Pl. 2: 1.

1998 *Eoplacognathus lindstroemi* (Hamar) – Bednarczyk, Pl. 2: 3, 5–7, 10.

Remarks. – *Eoplacognathus lindstroemi* was described and reconstructed by Bergström (1971).

Occurrence. – Andersön-B, -C. *P. anserinus-S. kielcensis* Subzone.

Material. – 10 Pa, 12 Pb.

Eoplacognathus reclinatus (Fåhræus, 1966)

Pl. 7: 1–5

Synonymy. –

Pb

1966 *Ambalodus reclinatus* n. sp. – Fåhræus, p. 19; Pl. 4: 3a–b.

Multielement

1971 *Eoplacognathus reclinatus* (Fåhræus) – Bergström, p. 139; Pl. 1: 11–13; Text-fig. 4–5: 7.

1973 *Eoplacognathus reclinatus* (Fåhræus) – Bergström, Fig. 2: 6.

1974 *Eoplacognathus reclinatus* (Fåhræus) – Viira, p. 81; Pl. 8: 12–13; Text-fig. 93–94.

1998 *Eoplacognathus reclinatus* (Fåhræus) – Bednarczyk, Pl. 2: 11.

Remarks. – The apparatus was reconstructed and described by Bergström (1971). Based on the present material, it may be added that the Pa elements have a short postero-lateral process (in terms of Bergström 1971), which makes an angle of 60–80° with the anterior process.

Occurrence. – Andersön-B. *P. serra - E. reclinatus* Subzone.

Material. – 4 Pa, 8 Pb.

Eoplacognathus suecicus Bergström, 1971

Pl. 7: 6–9, 13

Synonymy. –

Pa

1974 *Eoplacognathus suecicus* Bergström – Viira, p. 75; Pl. 8: 4–9; Text-figs. 82, 83.

Pb

1960 *Ambalodus* n. sp. 3 – Lindström, Fig. 5: 4, 9.

1967 *Ambalodus* n. sp. 3 Lindström – Viira, Fig. 3: 17a,b.

1972 *Eoplacognathus ambaloides* Viira – Viira, Fig. 6.

1974 *Eoplacognathus* aff. *foliaceus* (Fåhræus) – Viira, p. 78; Figs. 8, 9B, 87.

Multielement

1971 *Eoplacognathus suecicus* – Bergström, p. 141; Pl.1: 5–7.

1976 *Eoplacognathus suecicus* Bergström – Dzik, p. 409; Fig. 7.

? pt. 1978 *Eoplacognathus suecicus* Bergström – Löfgren, p. 59; Pl. 15: 9–11, 16–17 (only).

pt. 1978   *Eoplacognathus suecicus* Bergström – Löfgren,
              p. 59; Pl. 15: 18 (only).
1979   *Eoplacognathus suecicus* Bergström – Harris,
              Bergström, Ethington & Ross, Pl. 2: 8 – 10.
1985   *Eoplacognathus suecicus* Bergström – An, Du
              & Gao, Pl. 18: 1, 4.
? 1987   *Eoplacognathus suecicus* Bergström – Hün-
              icken & Ortega, p. 139; Pl. 7.1: 4 – 13.
pt. 1995a   *Eoplacognathus suecicus* Bergström – Lehnert,
              p. 87; Pl. 10: 10, *non* Pl. 10: 14.
1998b   *Eoplacognathus suecicus* Bergström – Zhang,
              pp. 70 – 71; Pl. 8: 11 – 13.

Remarks. – *Eoplacognathus suecicus* was described by Bergström (1971), Löfgren (1978) and Zhang (1998a, b), but their interpretations of *E. suecicus* are slightly different. Zhang (1998a, b) excluded the specimens pictured by Löfgren (1978, pl. 15: 9 – 14, 16 – 18) as *Eoplacognathus suecicus* from this species. Instead, they were referred to *Eoplacognathus pseudoplanus* (Viira). According to Zhang (1998b, p. 70) *E. suecicus* is distinguished from *E. pseudoplanus* by the "appearance of a secondary postero-lateral denticle row in the Pa elements". The Pa element figured by Löfgren (1978, pl. 15: 18) from her *E. suecicus* - *S. gracilis* Subzone, seems to have a secondary denticle row, which indicates that it actually belongs to *E. suecicus*. Therefore, it may be speculated that *E. pseudoplanus* and *E. suecicus* have overlapping ranges, and that *E. suecicus* has its first appearance in the Kundan *E. suecicus* - *S. gracilis* Subzone *sensu* Löfgren (1978) but becomes the dominant *Eoplacognathus* species in the slightly younger Aserian strata (the *E. suecicus* Zone *sensu* Zhang or the *E. suecicus* - *P. sulcatus* Subzone of Löfgren). The observations of *E. suecicus* within the present study are few but support this view, because *E. suecicus* has its first appearance with *Panderodus sulcatus* in the Andersön-B section but probably below this level at Steinsodden.

    The element figured by Lehnert (1995a, Pl. 10:14) is probably a *Polonodus* P element.

Occurrence. – Andersön-B, Steinsodden, Røste. *E. suecicus* Zone.

Material. – 19 Pa, 30 Pb.

Genus *Erraticodon* Dzik, 1978

Type species. – *Erraticodon balticus* Dzik, 1978

Erraticodon sp.

Pl. 7: 10 – 11, 14

Description. –

General remarks
The elements are hyaline. Both the cusp and the denticles are compressed and have sharp margins. The basal cavity is shallow and seems to extend to the tip of the processes. The few elements uncovered are broken which prevents a proper description.

P
The element is extensiform digyrate. The angle between the posterior and anterior process is 140°. Both the posterior and the anterior processes are denticulated. The cusp is recurved with a sharp anterior margin.

S
The element is bipennate with denticulated posterior and anterior processes. Cusp is recurved and denticles are reclined.

Remarks. – The material at hand is broken and does not represent the whole apparatus. It seems likely, however, that the species is separated from *E. balticus* Dzik 1978 and *Erraticodon* sp. Löfgren 1985 (p. 126, Fig. 4AT – AY) by having a more obtuse angle between the anterior and posterior processes in the P elements. Moreover, it lacks the extraordinarily large denticle on the posterior process typical of most elements in *E. balticus*.

Occurrence. – Andersön-B. *P. anserinus-S. kielcensis* Sub-zone.

Material. – 2 P, 1 S

Genus *Fahraeusodus* Stouge & Bagnoli, 1988

Type species. – ?*Microzarkodina adentata* McTavish, 1973

Remarks. – The apparatus comprises P, M and S elements (Stouge & Bagnoli 1988).

Fahraeusodus marathonensis (Bradshaw, 1969)

Pl. 7: 16

Synonymy. –

Sb
1969   *Gothodus marathonensis* n. sp. – Bradshaw, p. 1151;
              Pl. 137: 13 – 15; Text-figs. 3: S,T,U.

Sa
1969   *Roundya* sp. – Bradshaw, pp. 1160 – 1161; Pl. 137:
              17; Text-fig. 3A.

Sc

1969 *Paracordylodus* sp. – Bradshaw, p. 1159; Pl. 136: 12, 13.

Multielement

1989 *Microzarkodina? marathonensis* (Bradshaw) – Bauer, pp. 102–103; Fig. 4: 15, 20.

Description. –

All elements have small, lateral, median costae on the denticles.

P

The P element has an anterior, reclined cusp with medial lateral costae. The keels continues from the cusp to near the basal margin. Two large denticles with narrow medial costae are situated just behind the cusp. The posterior process is strongly flexed with 8 small denticles. The basal cavity is shallow with the apex situated beneath the posterior margin of the cusp.

S

The S elements were described by Bradshaw (1969).

Remarks. – The S elements obtained from the Stein section is similar to "*Microzarkodina*" *marathonensis* *sensu* Ethington & Clark (1982). On the other hand, the single element interpreted as the P element is clearly different from the "*M.*" *marathonensis* P element ("ozarkodiniform") of Ethington & Clark (1982; Pl. 5:14, 20) because it has much smaller denticles on the posterior process. In addition, the posterior process is twisted. This may indicate that "*M.*" *marathonensis* *sensu* Ethington & Clark instead belong to *F.* aff. *marathonesis*, which is the probable ancestor.

Occurrence. – Steinsodden. Interval from the upper part of the *B. norrlandicus - D. stougei* Zone to the basal part of the *B. medius - H. holodentata* Zone.

Material. – 1 P and 3 S elements.

Fahraeusodus aff. *marathonensis* (Bradshaw, 1969)

Pl. 7: 15

Synonymy. –

? 1982 "*Microzarhodina*" *marathonensis* (Bradshaw) – Ethington & Clark, pp. 55–56; Pl. 5: 14, 19, 20, 23, 24, 27.

1988 *Fahraeusodus marathonensis* (Bradshaw) – Bagnoli & Stouge, p. 119; Pl. 4: 15–17 (*cum. syn.*).

1991 *Fahraeusodus marathonensis* (Bradshaw) – Smith, pp. 36–37; Fig. 20a–g.

1995a *Fahraeusodus marathonensis* (Bradshaw) – Lehnert, p. 89; Pl. 7: 19, Pl. 8: 1.

Remarks. – All elements have small, lateral, median costae on the denticles, which is also a characteristic of *Fahraeusodus marathonensis* (Bradshaw) *sensu stricto*. *Fahraeusodus* aff. *marathonensis* was described by Stouge & Bagnoli (1988), who referred it to the *Fahraeusodus marathonensis*. The single specimen at hand is identical with the specimen referred to *Fahraeusodus marathonensis* P element by Stouge & Bagnoli (1988, p. 119, Pl. 4: 16).

Occurrence. – Andersön-B. *P. elegans* Zone.

Material. – One P element.

Genus *Goverdina* Fåhræus & Hunter, 1985

Type species. – *Goverdina alicula* Fåhræus & Hunter 1985.

Goverdina? sp.

Pl. 7: 17

Synonymy. –

pt. 1978 *Belodella jemtlandica* n. sp. – Löfgren, pp. 46–49; Pl. 15: 4 (only).

1984 *?Belodella* sp. A – Stouge, p. 61; Pl. 7: 15–16.

? 1985a *Goverdina alicula* n. gen. et sp. – Fåhræus & Hunter, p. 1178; Pl. 1: 6, 8; Pl. 2: 15–19; Text-fig. 3.

Description. – The element is plano-convex with 8–12 denticles. The cusp is proclined. The basal cavity is triangular with the apex placed near the anterior margin where the cusp meets the base. The element may be slightly inwardly bowed.

Remarks. – *Goverdina?* sp. co-occurs with *Ansella jemtlandica* (Löfgren) and it cannot be excluded that it forms a variation of the Sc element in this species. However, two important differences between *Goverdina?* sp. and typical *A. jemtlandica* were observed: the former develops a clearly more shallow basal cavity, and has relatively coarser (wider) denticles.

None of the other element types which characterises *Goverdina* Fåhræus & Hunter were recorded, and consequently the generic assignment is queried. The element was first described by Stouge (1984).

Occurrence. – Andersön-B. Interval from the upper part of the *B. medius - H. holodentata* Zone to lower part of the *P. graeai* Zone.

Material. – 6 plano-convex (Sc ?) elements.

Genus *Histiodella* Harris, 1962

Type species. – *Histiodella altifrons* Harris, 1962

Remarks. – The seximembrate *Histiodella* multielement apparatus was reconstructed independently by McHargue (1982) and Stouge (1984). Analysis made on a relatively large amount of material (1390 specimens) has shown that the ratio between P, S and M elements is about 20:4:1 (Stouge 1984).

Histiodella holodentata Ethington & Clark, 1982

Pl. 7: 18–19

Synonymy. –

1982	*Histiodella holodentata* n. sp. – Ethington & Clark, p. 47; Pl. 4: 1, 3, 4, 16.
? 1983	*Histiodella infrequensa* n. sp. – An *et al.*, Pl. 25: 1, 2.
1984	*Histiodella tableheadensis* n. sp. – Stouge, p. 87; Pl. 18: 8, 12–14; Text-fig. 17 (*cum. syn.*).
? 1985	*Histiodella intertexta* n. sp. – An *et al.*, Pl. 14: 15, 16.
pt. 1995a	*Histiodella holodentata* Ethington & Clark – Lehnert, pp. 90–91; Pl. 9: 10 (only).
1998b	*Histiodella tableheadensis* Stouge – Zhang, p. 72; Pl. 9: 14–15.

Description. – The P element is carminate. Commonly 5–8 denticles are present on both sides of the posteriorly placed cusp. The cusp and denticles are striated. The cusp is about twice as wide as the denticles. The length of elements varies between 0.25 and 0.43 mm in the present material. The stratigraphically older specimens often bear more slender denticles than the younger ones. The posterior margin of the P element makes an angle of 60–75° with the basal margin. A horizontal, median ledge divides the hyaline base from the upper, albid part of the element. Only P elements were observed in the present material.

The *H. holodentata* apparatus was described in detail by Ethington & Clark (1982) and Stouge (1984 [as *H. tableheadensis*]).

Remarks. – The *Histiodella* element figured by Lehnert (1995a, Pl. 9: 7) is characterised by relatively slender cusp, and does probably belong to *H. kristinae*. *H. holodentata* was reported from Hølonda in the Upper Allochthon of Norwegian Caledonides by Bergström (1979) (as *Histiodella* sp. cf. *serrata* Harris). The Hølonda Limestone is interpreted as an ocean island or island arc complex deposit, which palinspastically was situated close to Laurentia (e.g. Neumann & Bruton 1974).

The successor species *Histiodella kristinae* Stouge has been recorded from several areas on the Baltoscandian shelf (Lindström 1960; Viira 1974) and from Poland (Dzik 1976). This indicates that *Histiodella* preferred the distal, ocean–near parts of the Baltoscandian shelf during the relatively shallow water conditions that prevailed in the late Arenig and earliest Llanvirn. It became relatively common throughout the shelf during the succeeding early Llanvirn transgression.

Occurrence. – Steinsodden, Andersön-A, -C. *B. medius - H. holodentata* Zone – *P. graeai* Zone.

Material. – 37 P elements.

Plate VIII

1–3, 5: *Histiodella kristinae* Stouge.
1. P element, × 115. Sample 99479, Andersön-B. PMO 165.283.
2. P element, × 175. Sample 99479, Andersön-B. PMO 165.284.
3. P element, × 130. Sample 99479, Andersön-B. PMO 165.285.
5. P element, × 80. Sample 69668, Steinsodden. PMO 165.286.

6: *Histiodella* sp. A.
6. P element, × 210. Sample 97730, Jøronlia. PMO 165.287.

4, 7–8: *Lenodus* cf. *variabilis* (Sergeeva).
4. Pb element, × 80. Sample 99464, Andersön-B. PMO 165.288.
7. M element, × 120. Sample 69638, Steinsodden. PMO 165.289.
8. Sb element, × 95. Sample 99464, Andersön-B. PMO 165.290.

9–11: *Lenodus* sp. B.
9. Sd element, × 85. Sample 97831, Andersön-A. PMO 165.291.
10. Pb element, × 150. Sample 99462, Andersön-B. PMO 165.292.
11. Sd element, × 105. Sample 99462, Andersön-B. PMO 165.293.

12: *Lenodus* sp. A.
12. Sb element, × 95. Sample 99456, Andersön-B. PMO 165.294.

13: *Microzarkodina corpulenta* n. sp.
13. P element, × 80. Sample 69687, Steinsodden. PMO 165.295. Holotype.

14: *Microzarkodina flabellum* (Lindström).
14. P element, × 85. Sample 99456, Andersön-B. PMO 165.296.

15–20: *Microzarkodina parva* Lindström.
15. P element, × 120. Sample 97831, Andersön-A. PMO 165.297.
16. P element, × 100. Sample 97831, Andersön-A. PMO 165.298.
17. M element, × 100. Sample 97831, Andersön-A. PMO 165.299.
18. Sa element, × 90. Sample 97831, Andersön-A. PMO 165.300.
19. Sb element, × 120. Sample 97831, Andersön-A. PMO 165.301.
20. Sc element, × 120. Sample 97831, Andersön-A. PMO 165.302.

21: *Lenodus?* sp. C.
21. Pa element, × 30. Sample 97657, Glöte. PMO 165.303.

PLATE VIII

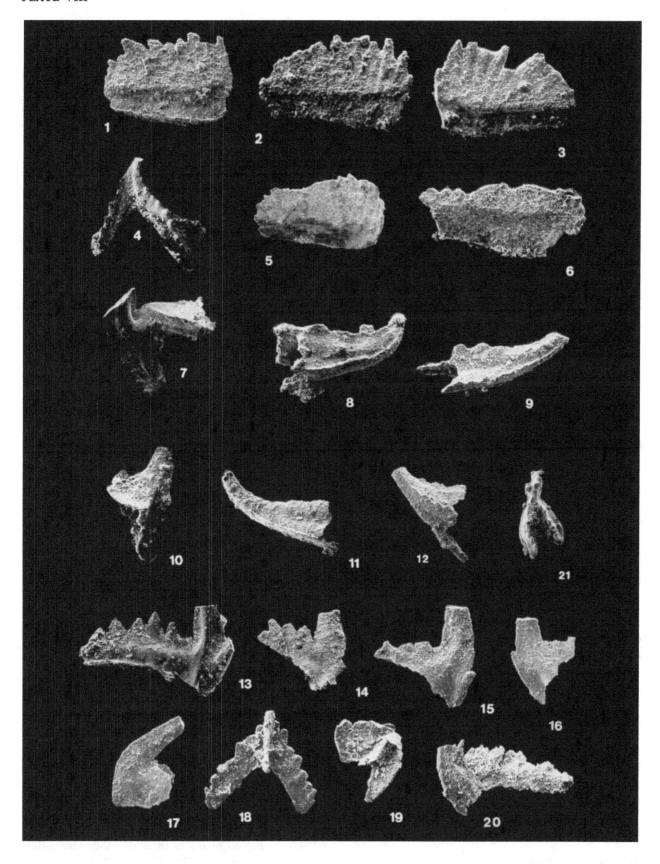

Histiodella kristinae Stouge, 1984

Pl. 8: 1–3, 5

Synonymy. –

1984   *Histiodella kristinae* n. sp. – Stouge, p. 87; Pl. 18: 1–7, 9–11; Text-fig. 17 (*cum. syn.*).

1994   *Histiodella kristinae* Stouge – Dzik, p. 110; Pl. 24: 28–30; Text-fig. 30.

1995a   *Histiodella kristinae* Stouge – Lehnert, p. 91; Pl. 9: 6, ?Pl. 10: 1.

? 1995a *Histiodella holodentata* Ethington & Clark – Lehnert, pp. 90–91; Pl. 9: 7.

1998b   *Histiodella kristinae* Stouge – Zhang, pp. 72–73; Pl. 9: 16–17.

Remarks. – The species was fully described by Stouge (1984). It can be added that the angle between the posterior and basal margins usually is between 50 and 60°, which is less than in *H. holodentata* (60–75°). The relatively slender cusp of the *Histiodella* element figured by Lehnert (1995a, Pl. 9: 7) makes it probable that it belong to *H. kristinae*.

Occurrence. – Steinsodden, Andersön-B. *P. graeai* Zone – *E. suecicus* Zone.

Material. – 131 P elements.

Histiodella sp. A

Pl. 8: 6

Synonymy. –

aff. 1979 *Histiodella* n. sp. 2 – Harris, Bergström, Ethington and Ross, Pl. 4: 12, 13.

Remarks. – The single element recorded here is a P element characterised by a median cusp with six reclined, posterior denticles and eight anterior denticles. The anterior denticles are proclined in the front and erect to reclined near the cusp. A horizontal, median ledge separates the hyaline base and the upper, albid part of the element. The reclined cusp is of similar size as the denticles. The element is distinguished from *Histiodella* n. sp. 2 Harris *et al.* 1979 by having a lower albid part of the element, a much narrower cusp, and an almost straight basal margin.

Occurrence. – Jøronlia. *P. graeai* Zone.

Material. – One P element.

Genus *Lenodus* Sergeeva, 1963

Type species. – *Lenodus clarus* Sergeeva, 1963

Remarks. – Before the multielement genus *Lenodus* was reconstructed by Stouge & Bagnoli (1990), *Lenodus* was considered as possibly unimembrate (Sergeeva 1963; Bergström in Robison 1981) or an element in the multielement genus *Amorphognathus* Sergeeva *sensu* Lindström (1977). Stouge & Bagnoli (1990) interpreted *Lenodus* as being septimembrate, comprising Pa, Pb, M, Sa, Sb, Sc and Sd elements and noted that it is distinguished from *Amorphognathus* by having mirror images of the left and right Pa elements (= Pb elements of Stouge & Bagnoli).

Lenodus cf. *variabilis* (Sergeeva, 1963)

Pl. 8: 4, 7–8

Synonymy. –

Pa

1960   *Amorphognathus* n. sp. 1 – Lindström, Text-fig. 4: 4.

cf. 1963   *Amorphognathus variabilis* n. sp. – Sergeeva, p. 105; Pl. 8: 15–17; Text-fig. 11.

non 1970 *Amorphognathus variabilis* Sergeeva – Fåhræus, p. 2065; Fig. 3E.

non 1970 *Amorphognathus variabilis* Sergeeva – Uyeno & Barnes, p. 106; Pl. 24: 5, 6.

1974   *Amorphognathus variabilis* Sergeeva – Viira, Pl. 7: 3, 4.

Pb

1960 *Ambalodus* n. sp. 2 – Lindström, Text-fig. 4: 5.

1963 *Ambalodus planus* n. sp. – Sergeeva, p. 105; Pl. 8: 11–15; Text-fig. 10.

1972 *Ambalodus planus* Sergeeva – Viira, pp. 45–46; Text-fig. 1.

1974 *Ambalodus planus* Sergeeva – Viira, pp. 53–54; Pl. 6: 22–24, 27, 30; textfigs. 40, 42.

Multielement

pt. 1976 *Amorphognathus variabilis* Sergeeva – Dzik, p. 432; Text-fig. 26a–e, g (only).

pt. 1978 *Eoplacognathus? variabilis* (Sergeeva) – Löfgren, pp. 57–59; Pl. 15: 15, 23–25, *non* Pl. 15: 22.

1983   *Amorphognathus variabilis* Sergeeva – Bergström, p. 38; Fig. 1.

non 1990 *Lenodus variabilis* (Sergeeva) – Dzik, Fig. 9.

1992   *Eoplacognathus? variabilis* (Sergeeva) – Rasmussen, pp. 278–279; Pl. 6Q–U.

pt. 1994 *Lenodus variabilis* (Sergeeva) – Dzik, pp. 86–88; Pl. 19:16–23, Pl. 20: 1, 3, 4, 6–8 (only).

cf. 1998b *Lenodus variabilis* (Sergeeva) – Zhang, pp. 74–75; Pl. 11: 1–10.

? 1998b *Yangtzeplacognathus crassus* (Chen & Zhang) – Zhang, pp. 96; Pl. 20: 5–8.

Comments to the synonymy list. – The holotype of *Lenodus variabilis* (i.e. *Amorphognathus variabilis* Sergeeva 1963, Pl. 8: 16) forms an almost symmetrical cross with a central cusp. Thus, it differs from the corresponding *L.* cf. *variabilis* elements (Pa-1), which makes an asymmetrical cross with lateral processes that are not directly jointed. The Pa element figured by Fåhræus (1970, Fig. 3E) is *Polonodus? tablepointensis* Stouge. The apparatus referred to *Lenodus variabilis* by Dzik (1990, Fig. 9) includes a Pa element which has a lateral expansion from the posterior process. This expansion is uncommon in elements referred to *L.* cf. *variabilis* in this paper but is typical for Pa elements included in *Eoplacognathus suecicus* Bergström. The M and S elements figured by Dzik (1990, Fig. 9) are typical *Lenodus* elements, however. The dextral Pb element depicted by Löfgren (1978; Pl. 15: 22) differs from the Pb element in *L.* cf. *variabilis* by having a smaller angle between the anterior and lateral process. The oz elements figured by Dzik (1994, Pl. 20: 2, 5) are here interpreted as Pb elements of *Polonodus*.

Description. –

Pa, type 1

The elements are stelliplanate and forms an asymmetrical cross, which are characterised by four denticulated, approximately equally long processes. The posterior and postero-lateral processes are longer and wider than the anterior and antero-lateral processes. The denticles are situated in a median row. The cusp is short. All element types have a deep and wide basal cavity.

Pa, type 2

The elements are stelliplanate with four denticulated processes. The Pa type 2 element is distinguished from the type 1 element by having an anterior process that is almost perpendicular to both the lateral and posterior process. The posterior and postero-lateral processes are clearly longer than the additional processes. Sinistral and dextral elements are mirror images.

Sinistral Pb

Sinistral Pb elements are pastiniplanate, commonly nearly T-shaped but sometimes more like a Y, and with three denticulated processes of subequal length. The anterior process may be slightly longer than the others. The platform-ledges of the posterior and lateral processes are wider than that of the anterior process.

Seen from above, the posterior and lateral processes makes an almost straight line while the slender, anterior process makes an angle of 90–100° with this line. The denticulation on the posterior process is almost median, whereas it is placed close to the outer edge in dextral Pb elements.

Dextral Pb

The elements are pastiniplanate, T-shaped, with three processes of subequal length. Seen from above, the anterior and posterior process makes an almost straight line, although the anterior process may curve slightly away from the lateral process. The cusp is characterised by a carina which reach the lateral process.

M

The element is geniculate with an anterior, denticulated process and an inner-lateral, denticulated or nondenticualated process. The inner-lateral process is shorter or of the same size as the posterior and anterior processes. A large basal sheath occurs between the inner-lateral process and the posterior process. The angle between the anterior and posterior processes varies between 60° and 80° in a lateral view. The cusp is erect or recurved.

Sa

The Sa element is a an alate ramiform, bearing one posterior and two lateral processes, all of which are denticulated. The element is characterised by a large basal sheath. The cusp is proclined.

Sb

The element is tertiopedate and forms an angle of 25–30° between the anterior and posterior processes. Processes are denticulated. The element is characterised by a large basal sheet which extends almost to the tips of the processes. The cusp is proclined.

Sc

The Sc element is asymmetrically tertiopedate with denticulated anterior and posterior processes. The anterior process is twisted and the denticles point upwards and inwards. Generally, the angle between the two processes is 30–45°. The lateral process-like extension is very short and undenticulated. The cusp is proclined.

Sd

The Sd element is asymmetrically quadriramate with four denticulated processes. The element is characterised by a large, distally prolonged basal sheath. The angle between the individual processes is 20–30°. The cusp is proclined.

Remarks. – The M element is similar to the *Lenodus* n. sp. A element illustrated by Stouge & Bagnoli (1990, Pl. 4: 9) but this species includes some very elongate Pa-2 elements, which were not observed in the present material. However, the specimens at hand are not well-preserved, and it is likely that *L.* cf. *variabilis* in this study includes more than one species, among others *Eoplacognathus crassus* Chen & Zhang 1993 (Zhang 1997, 1998b). The S elements are similar to the elements figured as *Lenodus* sp. A by Stouge & Bagnoli (1990, pp. 18–19; Pl. 4: 10–13) and as *Amorphognathus* elements by Löfgren (1990, p. 250; Fig. 1t–ac).

Occurrence. – Andersön-A, -B, -C, Steinsodden, Røste, Jøronlia. *B. norrlandicus* – *D. stougei* Zone – *E. suecicus* Zone.

Material. – 23 Pa, 58 Pb, 10 M, 6 Sa, 9 Sb, 2 Sc, 13 Sd.

Lenodus sp. A

Pl. 8: 12

Description. –

Dextral Pb
The element is pastinate with denticulated anterior and posterior processes. The processes are slender and blade-like, and lack platform-ledges. Denticles are fused. The angle between the anterior and posterior process is about 85° in lateral view. The lateral process is broken close to the cusp in all specimens. The cusp is suberect or slightly proclined. A wide basal sheet is present. The basal cavity is wide with the apex situated close to the anterior margin of the cusp.

Sb
The element is asymmetrical tertiopedate and forms an angle of 30–40° between the anterior and posterior processes. The processes are denticulated. The element is characterised by a large basal sheet which extends almost to the tip of the processes. The cusp is proclined.

Remarks. – The Pb element lacks the platform-ledges which typifies *Lenodus* cf. *variabilis*.

Occurrence. – Andersön-A, -B. *P. rectus* – *M. parva* Zone.

Material. – 3 dextral Pb, 7 Sb.

Lenodus sp. B

Pl. 8: 9–11

Description. –

Dextral Pb
The single Pb element is broken and lacks the lateral

process and most of the posterior process. The general outline, however, is similar to *Lenodus* sp. A (see description of this), except that the angle between the anterior and posterior processes is more than 90° in lateral view (Pl. 8: 10).

Sa
The Sa element is alate ramiform carrying one posterior and two lateral processes, all of which are directed to the posterior end. The processes are denticulated or develop narrow edges. The basal sheath is large. The cusp is proclined.

Sb
The element is tertiopedate ramiform and forms an angle of 25–30° between the anterior and posterior processes. Processes are denticulated. The element is characterised by a large basal sheet which extends almost to the tip of the processes. The cusp is proclined.

Sc
The Sc element is asymmetrically tertiopedate with denticulated anterior and posterior processes. The anterior process is twisted and the denticles point upwards and inwards. Generally, the angle between the two processes is 30–52°. The lateral extension is very short and undenticulated. The extension is process-like in some specimens whereas it only forms a small, oblong knob in others. The cusp is proclined.

Sd
The Sd element is asymmetrically quadriramate with four denticulated processes which are posteriorly directed. The element is characterised by a large basal sheath. The angle between the anterior and posterior processes is 20–30°. The cusp is proclined.

Remarks. – No pastiniplanate elements have been observed.

Occurrence. – Andersön-A, -B. *B. norrlandicus* – *D. stougei* Zone.

Material. – 1 Pb, 8 Sa, 1 Sb, 2 Sc, 5 Sd.

Lenodus? sp. C

Pl. 8: 21

Description. –

Pa
The element is stelliplanate pectiniform and develops four denticulated processes. The posterior and postero-lateral processes are larger than the anterior and

antero-lateral processes. The blade-like, anterior process and the proximal part of the postero-lateral process form a straight line, whereas the distal part of the postero-lateral process is bent 35° away from the line and is almost parallel to the posterior process. The denticles are situated in a median row. The cusp is short. The basal cavity is relatively deep and wide.

Pb

The Pb elements are pastiniplanate pectiniform carrying three denticulated processes. Both specimens at hand are fragmented. The best preserved specimen is a Y-shaped sinistral element, characterised by a slender, almost blade-like anterior process and a relatively wide posterior process. The angle between the two process is 150°. The lateral process is lacking. The basal cavity is relatively wide.

Remarks. – The Pa element is distinct by its posterior and postero-lateral processes that are bent in the same direction. Dzik (1976, 1989) showed that the feature is also known from *Saggitodontina kielcensis* but that species is characterised by clearly wider platform-ledges on the postero-lateral process. The Pa element shares also characteristics with the stelliplanate elements of *Eoplacognathus suecicus* Bergström, but because the single element at hand is broken the name has been put into open nomenclature.

Occurrence. – Glöte. Uncertain stratigraphical level between the *B. medius – H. holodentata* Zone and *E. suecicus* Zone.

Material. – 1 Pa, 2 Pb.

Genus *Lundodus* Bagnoli & Stouge, 1997

Type species. – *Acodus gladiatus* Lindström, 1955

Remarks. – The multielement genus *Lundodus* was reconstructed by Bagnoli and Stouge (1997). The genus is basically bimembrate comprising P and S elements.

Lundodus gladiatus (Lindström, 1955)

Pl. 1: 2

Synonymy. –

1955a *Acodus gladiatus* n. sp. – Lindström, pp. 544 – 545; Pl. 3: 10 – 12.

1997 *Lundodus gladiatus* (Lindström) – Bagnoli & Stouge, p. 146; Pl. 3: 13 – 16 (*cum. syn.*).

Remarks. – The P elements were described as the formspecies *Acodus gladiatus* by Lindström (1955a), and the S elements as "*Belodella*" sp. B by Serpagli (1974, p. 39, Pl. 7: 1a – c, Pl. 20: 11).

Only the P elements have been observed in this study.

Occurrence. – Andersön-A, *O. evae* Zone.

Material. – 4 specimens.

Genus *Microzarkodina* Lindström, 1971

Type species. – *Prioniodina flabellum* Lindström, 1955

Remarks. – The apparatus is seximembrate comprising P, M, Sa, Sb, Sc and Sd elements. The complete multielemental apparatus was illustrated by Van Wamel (1974).

Microzarkodina corpulenta n. sp.

Pl. 8: 13

Derivation of name. – Corpulentus (Latin); fleshy, fat.

Type-locality. – Steinsodden, Moelv, Norway

Type-stratum. – The lowermost 20 cm of the Steinsholmen Member, sample 69687, *Microzarkodina flabellum – D. forceps* Zone.

Holotype. – Repository PMO 165.295. P element (Plate 8: 13).

Diagnosis. – A species of *Microzarkodina* characterised by a P element with thickened ledges on the lateral surfaces of the posterior process. The length in anterior-posterior direction is always larger than the distance from the tip of the cusp to the basal margin.

Description. – *Microzarkodina corpulenta* n. sp. differs from *M. flabellum* solely by the distinct morphology of the P element. In sample 69687 where *M. corpulenta* and *M. flabellum* occur together, the M and S elements have been divided proportionally between the two species.

P

The P element is carminate pectiniform with thickened lateral surfaces of the denticulated, posterior process. The thickening is most conspicuous proximally. A large, slightly twisted denticle is situated in front of the cusp. Measurements made on the small amount of material at hand show that the length/height ratio varies between 1.28 and 1.45 and that the angle between the cusp and the anterior denticle is 38–54°. Accordingly, the length/height ratio is significantly larger than in *M. flabellum* (0.7–1.1), while the cusp-denticle angle is about the same. The denticles are generally more wide and stout compared with the cusp than is typical for other *Microzarkodina* species. The cusp may usually (but not always) carry a median lateral costa on one or both sides (see Pl. 8: 13). The basal cavity shallows gradually towards the distal part of the posterior process.

M

The element is geniculate coniform with a posteriorly extended base. The basal cavity is wide in the proximal part of the element and makes a narrow groove distally. Cusp is reclined with sharp edges at the anterior and posterior margins and a distinct carina on both lateral surfaces.

Sa

Sa elements are almost symmetrical, alate ramiforms with long, recurved, denticulated, lateral processes. The posterior keel continues in the downward, posterior direction and forms a minute adenticulated extension. Both lateral processes are slightly twisted and bear fused denticles that become smaller in the distal parts of the processes. Basal cavity is shallow and forms a narrow groove beneath the lateral processes. The cusp is proclined.

Sc

The element is dolabrate ramiform with a long, bowed posterior process. It has commonly more than 13 reclined, fused denticles. The cusp varies from proclined to recurved with a carina on both lateral surfaces. The basal cavity continues as a narrow groove through the entire process.

Sd

The Sd element is quadriramate *sensu lato* as the anterior, posterior, and inner-lateral processes are strongly reduced. The outer-lateral process is long, weakly curved inward and usually has more than twelve fused, reclined denticles. The cusp is proclined. The basal cavity continues as a narrow groove through the entire process. The anterior and posterior processes are adenticulated, whereas the inner-lateral process typically bears two denticles.

Remarks. – In P elements, both the basal cavity and the basal sheath are more reduced than in other *Microzarkodina* species hitherto described. Because the distribution of the S and M elements was based on the relative number of P elements of *M. corpulenta* and *M. flabellum* (in sample 69687), no Sb elements were referred to *M. corpulenta*. However, there is no reason to doubt the existence of a Sb element in the *M. corpulenta* apparatus.

Occurrence. – Steinsodden. *M. flabellum – D. forceps* Zone.

Material. – 16 P, 2 M, 1Sa, 1Sc, 1Sd.

Microzarkodina flabellum (Lindström, 1955)

Pl. 8: 14

Plate IX

1–6: *Microzarkodina ozarkodella* Lindström.
1. P element, ×55. Sample 99471, Andersön-B. PMO 165.304.
2. P element, ×65. Sample 99471, Andersön-B. PMO 165.305.
3. M element, ×85. Sample 99471, Andersön-B. PMO 165.306.
4. Sa element, ×55. Sample 99471, Andersön-B. PMO 165.307.
5. Sc element, ×65. Sample 99471, Andersön-B. PMO 165.308.
6. Sd element, ×80. Sample 99471, Andersön-B. PMO 165.309.

7: *Microzarkodina hagetiana* Stouge & Bagnoli.
7. P element, ×120. Sample 97831, Andersön-A. PMO 165.310.

8–20: *Minimodus poulseni* gen. et sp. nov.
8. Drepanodontiform 1 element, ×70. Sample 99478, Andersön-B. PMO 165.311.
9. Drepanodontiform 1 element, ×50. Sample 99478, Andersön-B. PMO 165.312.
10. Drepanodontiform 1 element, ×60. Sample 99478, Andersön-B. PMO 165.313.
11. Drepanodontiform 2 element, ×80. Sample 99478, Andersön-B. PMO 165.314.
12. Drepanodontiform 2 element, ×80. Sample 99478, Andersön-B. PMO 165.315.
13. Drepanodontiform 3 element, ×85. Sample 99478, Andersön-B. PMO 165.316.
14. Drepanodontiform 3 element, ×35. Sample 99478, Andersön-B. PMO 165.317.
15. Drepanodontiform 3 element, ×65. Sample 99478, Andersön-B. PMO 165.318.
16. Scandodontiform element, ×95. Sample 99479, Andersön-B. PMO 165.319. Holotype.
17. Oistodontiform element, ×55. Sample 99478, Andersön-B. PMO 165.320.
18. Oistodontiform element, ×45. Sample 99478, Andersön-B. PMO 165.321.
19. Scandodontiform element, ×80. Posterior view. Sample 99479, Andersön-B. PMO 165.322.
20. Same specimen as in Pl. 9:19. Lateral view. ×100.

PLATE IX

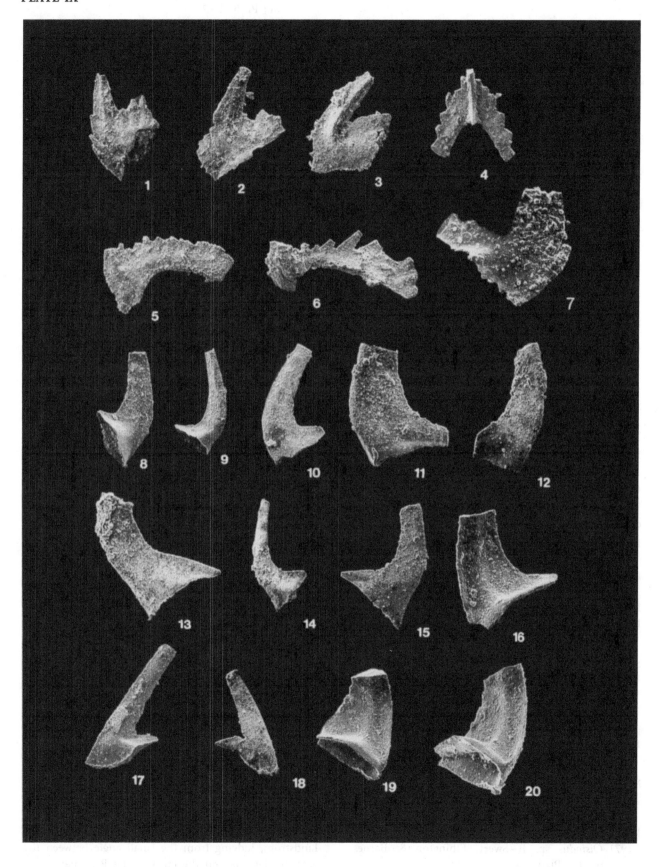

Synonymy. –

P

1955a   *Prioniodina flabellum* n. sp. – Lindström, p. 587;
Pl. 6: 23 – 25.

1960   *Prioniodina flabellum* Lindström – Lindström,
p. 90; Figs. 3: 10, 4: 12.

1961   *Prioniodina* cf. *flabellum* Lindström – Wolska,
p. 354; Pl. 4: 4a – b.

1964   *Ozarkodina flabellum* (Lindström) – Lindström,
Fig. 10P.

1967   *Prioniodina flabellum* Lindström – Viira, Fig. 1: 28.

? 1971   *Prioniodina? flabellum* Lindström? – Sweet,
Ethington & Barnes, p. 168; Pl. 1: 12.

1974   *Prioniodina flabellum* Lindström – Viira, p. 31;
Pl. 5: 3 – 4; Text-fig. 14a,d.

Sa

1955a   *Trichonodella alae* n. sp. – Lindström, p. 599; Pl.
6: 38 – 40.

1960   *Trichonodella alae* Lindström – Lindström, Fig. 3: 5.

1964   *Trichonodella alae* (trichonodelliform element)
Lindström – Lindström, p. 87; Fig. 31: L.

1967   *Trichonodella alae* Lindström – Viira, Fig. 1: 19.

1974   *Trichonodella alae* Lindström – Viira, p. 22, 127;
Pl. 5: 5.

Sb

1955a   *Trichonodella? irregularis* n. sp. – Lindström,
p. 600; Pl. 6: 21 – 22.

1964   *Trichonodella alae* (oulodiform element) Lindström – Lindström, p. 87; Fig. 31N.

1974   *Trichonodella? irregularis* Lindström – Viira, p.
127; Pl. 5: 6.

Sc

1955a   *Cordylodus perlongus* n. sp. – Lindström, p. 552;
Pl. 6: 36 – 37.

1960   *Cordylodus perlongus* Lindström – Lindström,
Fig. 3: 1.

1967   *Cordylodus perlongus* Lindström – Viira, Fig. 1: 21.

1974   *Cordylodus perlongus* Lindström – Viira, pp. 22,
63; Pl. 5: 7 – 8.

Sd

1960   *Trichonodella? irregularis* Lindström – Lindström,
Fig. 3: 2.

1964   *Trichonodella alae* (sannemanulla element) – Lindström, p. 87; Fig. 31M.

1967   *Trichonodella? irregularis* Lindström – Viira,
Fig. 1: 20.

M

1955a   *Oistodus linguatus complanatus* n. sp. – Lindström, p. 578; Pl. 3: 37 – 38.

1967   *Oistodus complanatus* Lindström – Viira, Fig. 1: 5.

? 1971   *Oistodus* sp. B – Sweet, Ethington & Barnes,
p. 168; Pl. 1: 13.

1974   *Oistodus complanatus* Lindström – Viira, pp. 22,
96; Pl. 5: 1 – 2.

Multielement

1971   *Microzarkodina flabellum* (Lindström) –
Lind-ström, p. 58; Pl. 1: 6 – 11; Text-fig. 19,20.

1974   *Microzarkodina flabellum* (Lindström) – van
Wamel, p. 70; Pl. 7: 18 – 23.

non 1977   *Microzarkodina flabellum* (Lindström) –
Gedik, p. 42; Pl. 2: 22 – 24.

pt. 1978   *Microzarkodina flabellum* (Lindström) –
Löfgren, p. 61; Pl. 11: 27 – 36; Text-fig.
27A, *non* Fig. 27B.

? 1981   *Microzarkodina flabellum* (Lindström) –
Nowlan, p. 14; Pl. 2: 1 – 5.

1981   *Microzarkodina flabellum* (Lindström) –
Ethington & Clark, p. 54; Pl. 4: 2, Pl. 5:
21, 22, 25, 26.

non 1981   *Microzarkodina flabellum* (Lindström) –
Cooper, p. 171; Pl. 28: 2 – 4.

non 1985   *Microzarkodina flabellum* (Lindström) – An,
Du & Gao, Pl. 15: 5 – 10.

1985   *Microzarkodina flabellum flabellum* (Lindström) – Löfgren, p. 127; Fig. 4H – N.

1990   *Microzarkodina flabellum* (Lindström) –
Stouge & Bagnoli, pp. 19 – 20; Pl. 6: 1 – 7.

1990   *Microzarkodina flabellum* (Lindström) –
Dzik, Fig. 5B.

1991   *Microzarkodina flabellum* (Lindström) –
Rasmussen, p. 279; Figs. 7A – D, F.

1993b   *Microzarkodina flabellum flabellum* (Lindström) – Löfgren, Fig. 6Q – U, Y, ?Z.

1994   *Microzarkodina flabellum flabellum*
(Lindström) – Löfgren, Fig. 8: 43, 44.

non 1995a   *Microzarkodina flabellum* (Lindström) –
Lehnert, p. 97; Pl. 6: 16 – 23.

1995a   *Microzarkodina flabellum* (Lindström) –
Löfgren, Fig. 7i – o.

1998   *Microzarkodina flabellum* (Lindström) –
Bednar-czyk, Pl. 1: 3, 8, 16.

Comments to the synonymy list. – *Oistodus* sp. B of
Sweet, Ethington & Barnes (1971; Pl. 1: 13),
Prioniodina? flabellum? of Sweet, Ethington & Barnes
(1971; Pl. 1: 12), and the specimens referred to
Microzarkodina flabellum by Ethington & Clark (1981;
Pl. 4: 2, Pl. 5: 21, 22, 25, 26) and Nowlan (1981; Pl. 2:
1 – 5) presumably belong to *Microzarkodina*, but it is
not possible to identify these specimens to the species
level based on the illustrated material.

The elements illustrated by Gedik (1977; Pl. 2: 22 –
24), Löfgren (1978, Text-fig. 27B), and Lehnert (1995a;
Pl. 6: 16 – 23) belong to *Microzarkodina parva*
Lindström, judging from the small angle between the
anterior denticle and the cusp of the P element.

The specimens presented by Cooper (1981; Pl. 28: 2–4) have a denticulation markedly different from *Micro-zarkodina* and belongs probably to a separate genus.

An, Du & Gao (1985; Pl. 15: 5–10) presented a complete *Microzarkodina* apparatus except for the Sd element. The P element had a small angle (less than 35°) between the anterior denticle and the cusp, which is typical of *M. parva*.

Description. – The individual elements were described by Lindström (1955a). The following characteristics may be added to the description of the P element: P element is carminate to angulate with a denticulated posterior process and a short anterior process-like extension bearing one denticle. This denticle is commonly twisted and in lateral view it makes an angle of about 40–50 (60)° with the cusp. The length/height ratio is between 0.7 and 1.1. The description of the M and S elements fits the description of the corresponding elements in *M. corpulenta*.

Remarks. – *M. flabellum* may be separated from *M. parva* on two characters: 1) The P element of *M. parva* is "distinguished by the fact that denticles of the ozarkodiniform element reach considerably less than the height of the cusp" (Lindström 1971, p. 59); and 2) the angle between the anterior margin of the cusp and the anterior margin of the anterior denticle is larger in *M. flabellum* than in *M. parva* (Löfgren, 1985). Rasmussen (1991) noted that this angle is less than 36° in *M. parva* while it is more than 36° in *M. flabellum*.

Occurrence. – Andersön-A, -B, Herram, Steinsodden. *M. flabellum* – *D. forceps* Zone – *P. rectus* – *M. parva* Zone.

Material. – 138 P, 39 M, 25 Sa, 21 Sb, 44 Sc, 15 Sd.

Microzarkodina hagetiana Stouge & Bagnoli, 1990

Pl. 9: 7

Synonymy. –

1990 *Microzarkodina hagetiana* n. sp. – Stouge & Bagnoli, p. 20; Pl. 6: 17–24 (*cum. syn.*).
pt. 1991 *Microzarkodina parva* Lindström – Rasmussen, p. 279; Fig. 7H, K (only).
1998b *Microzarkodina hagetiana* Stouge & Bagnoli – Zhang, p. 76; Pl. 12: 7–10.

Remarks. – *M. hagetiana* was fully described by Stouge & Bagnoli (1990).

Occurrence. – Andersön-A, -B, Steinsodden, Røste, Haugnes. *B. medius* – *H. holodentata* Zone – *M. ozarkodella* Zone.

Material. – 136 P, 111 M, 17 Sa, 7 Sb, 44 Sc, 9 Sd.

Microzarkodina ozarkodella Lindström, 1971

Pl. 9: 1–6

Synonymy. –

P

1960 Prioniodina n. sp. 1 – Lindström, Fig. 5: 1.
1960 *Prioniodina* n. sp. 2 – Lindström, Fig. 5: 2.
1967 Prioniodina sp. 1 Lindström – Viira, Fig. 1: 29.
1967 *Prioniodina* sp. 2 Lindström – Viira, Fig. 1: 30.
pt. 1974 *Prioniodina* sp. 1 + 2 Lindström – Viira, p. 31; Pl. 5: 34–38; Text-fig. 14, (only).
1974 *Prioniodina* sp. – Viira, Fig. 14.

Multielement

1971 *Microzarkodina ozarkodella* n. sp. – Lindström, p. 59; Pl. 1: 15–17.
1976 *Microzarkodina ozarkodella* Lindström – Dzik, Fig. 35i–l.
1978 *Microzarkodina ozarkodella* Lindström – Löfgren, p. 62; Pl. 11: 37–47; Text-fig. 27C–G.
1994 *Microzarkodina ozarkodella* Lindström – Dzik, p. 113; Pl. 24: 5–9; Text-fig. 31c.
1998 *Microzarkodina ozarkodella* Lindström – Bednarczyk, Pl. 1: 17.
1998b *Microzarkodina ozarkodella* Lindström – Zhang, pp. 76–77; Pl. 12: 11–13.

Remarks. – P element is angulate. *M. ozarkodella* is distinguished by having two or more denticles in front of the cusp. The differences between *M. ozarkodella* and *M. hagetiana* are negligible in the M, Sa, Sb, Sc and Sd elements. These elements agree with the description of *M. hagetiana* by Stouge & Bagnoli (1990). Löfgren (1985, p. 123) noted that P elements of *Microzarkodina* with more than one anterior denticle occur in very low frequencies as early as the *B. navis* Zone, but the elements were considered as abnormal *M. flabellum* elements.

Occurrence. – Andersön-A, -B, -C, Steinsodden, Røste. *B. medius* – *H. holodentata* Zone – *E. suecicus* Zone.

Material. – 114 P, 75 M, 9 Sa, 8 Sb, 46 Sc, 28 Sd.

Microzarkodina parva Lindström, 1971

Pl. 8: 15–20

1971   *Microzarkodina parva* n. sp. – Lindström, p. 59;
       Pl. 1: 14.
1976   *Microzarkodina   parva*   Lindström – Dzik,
       Fig. 35a – h.
1977   *Microzarkodina flabellum* (Lindström) – Gedik,
       p. 42; Pl. 2: 22 – 24.
pt. 1978 *Microzarkodina flabellum* (Lindström) – Löf-
       gren, p. 61; Fig. 27B (only).
? 1981   *Microzarkodina   flabellum   (Lindström) –*
       Nowlan, p. 14; Pl. 2: 1 – 5.
1985   *Microzarkodina flabellum parva* Lindström –
       Löfgren, p. 127; Fig. 4A – G.
pt. 1991 *Microzarkodina parva* Lindström – Rasmussen,
       p. 279; Fig. 7E, G – K (only).
1994   *Microzarkodina flabellum parva* Lindström –
       Löfgren, Fig. 8: 45, 46.
1995a *Microzarkodina flabellum* (Lindström) – Leh-
       nert, p. 97; Pl. 6: 16 – 23.
1995b *Microzarkodina   parva*   Lindström – Löfgren,
       Fig. 9: o – v.
1998b *Microzarkodina   parva*   Lindström – Zhang,
       pp. 75 – 76; Pl. 12: 1 – 6.

Remarks. – See Lindström (1971) for description. Stouge
& Bagnoli (1990) revised *Microzarkodina parva* and
included Kundan specimens, which were earlier referred
to *M. parva* in the new species *M. hagetiana* (Stouge &
Bagnoli 1990).

Occurrence. – Andersön-A, -B, Steinsodden, Haugnes,
Hestekinn. *P. rectus – M. parva Zone – B. medius – H.
holodentata* Zone.

Material. – 103 P, 65 M, 14 Sa, 6 Sb, 28 Sc, 14 Sd.

Genus *Minimodus* n. gen.

Type species. – *Minimodus poulseni* n. sp.

Derivation of name. – After *minimus* (Latin), smallest.
Refers to the shallow basal cavity.

Diagnosis. – Apparatus is quinquemembrate with laterally compressed scandodontiform, oistodontiform and
three types of drepanodontiform elements. Elements are
characterised by a shallow basal cavity and keel on the
cusp and the upper margin.

Remarks. – *Minimodus* n. gen. includes some of the
morphotypes which were incorporated in *Parapaltodus*

Stouge by Stouge (1984). However, *Parapaltodus simplicissimus* Stouge, which is the holotype of *Parapaltodus*, is
bimembrate and comprises scandodontiform and drepanodontiform elements with a characteristic, very short
upper margin (Stouge 1984, p. 48). No such elements were
associated with *Minimodus* n. gen., so the validity of
Parapaltodus is recognised.

The apparatus composition of *Minimodus* n. gen.
resembles that of *Paltodus* Pander 1856, but the latter
lacks the scandodontiform element. In addition,
Paltodus Pander *sensu* Lindström (1971) lacks elements
with a proclined cusp, while this characterises the
drepanodontiform elements of *Minimodus*.

Minimodus poulseni n. sp.

Pl. 9: 8 – 20

Synonymy. –

Drepanodontiform 1
pt. 1984 *Paltodus?* cf. *jemtlandicus* Löfgren – Stouge,
       p. 56; Pl. 4: 28 (only).
? 1994   *Tripodus?* sp. – Dzik, p. 79; Pl. 24: 27 (only).

Drepanodontiform 2
pt. 1984 *Parapaltodus angulatus* (Bradshaw) – Stouge,
       pp. 47 – 48; Pl. 2: 2 (only).

Plate X

1–13: *Nordiora torpensis* gen. et sp. nov.
 1. Pa element, × 100. Sample 99465, Andersön-B. PMO 165.323
 2. ?P element, × 80. Sample 97715, Hestekinn. PMO 165.324.
 3. Pb element (outer lateral view), × 70. Sample 97763, Røste. PMO
    165.325.
 4. Pb element (inner lateral view), × 50. Sample 97763, Røste. PMO
    165.326.
 5. Pb element (inner lateral view), × 100. Sample 97763, Røste. PMO
    165.327.
 6. M element, × 100. Sample 97763, Røste. PMO 165.328.
 7. M element, × 55. Sample 97763, Røste. PMO 165.329.
 8. Sa element, × 100. Sample 97763, Røste. PMO 165.330.
 9. Sb element (outer lateral view), × 65. Sample 97763, Røste. PMO
    165.331.
10. Sb element (inner lateral view), × 60. Sample 97763, Røste. PMO
    165.332.
11. Sc element, × 60. Sample 97763, Røste. PMO 165.333.
12. Sd element (outer lateral view), × 100. Sample 97763, Røste. PMO
    165.334. Holotype.
13. Sd element (inner lateral view), × 65. Sample 97763, Røste. PMO
    165.335.

14–18: *Nordiora* sp. A.
14. Pa element, × 150. Sample 99471, Andersön-B. PMO 165.336.
15. M element, × 70. Sample 99471, Andersön-B. PMO 165.337.
16. Sa element, × 95. Sample 99471, Andersön-B. PMO 165.338.
17. Sb element, × 95. Sample 99471, Andersön-B. PMO 165.339.
18. Sb element, × 95. Sample 99471, Andersön-B. PMO 165.340.

PLATE X

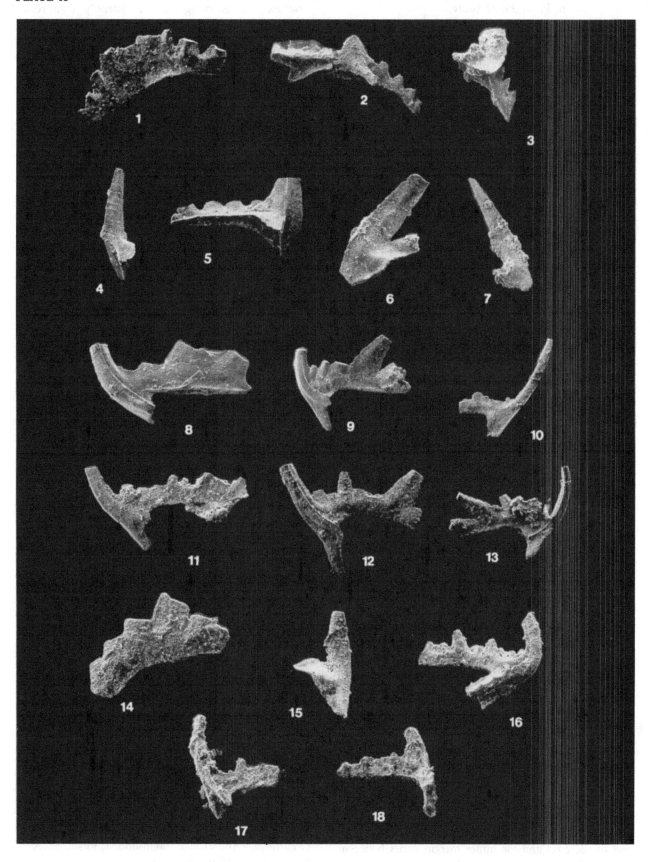

pt. 1984 *Paltodus?* cf. *jemtlandicus* Löfgren – Stouge,
　　p. 56; Pl. 4: 29, 32 (only).

Drepanodontiform 3
pt. 1984 *Paltodus?* cf. *jemtlandicus* Löfgren – Stouge,
　　p. 56; Pl. 4: 27 (only).
pt. 1984 *Parapaltodus angulatus* (Bradshaw) – Stouge,
　　pp. 47 – 48; Pl. 2: 1 (only).
pt. 1994 *Tripodus?* sp. – Dzik, p. 79; Pl. 24: 26 (only).

Oistodontiform
1978　　　　Gen. et sp. indet. A – Löfgren, p. 118, Pl. 4: 12.
pt. 1984　　*Paltodus?* cf. *jemtlandicus* Löfgren – Stouge,
　　　　　　p. 56; Pl. 4: 26, 30, 31, 33 (only).
? pt. 1998b *Triangulodus maocaopus* n. sp. – Zhang,
　　　　　　pp. 94 – 95; Pl. 18: 9 (only).

Derivation of name. – In the honour of Dr. phil.
Valdemar Poulsen, Professor at the Dept. of Historical
Geology & Palaeontology, University of Copenhagen,
Denmark.

Type locality. – The island Andersön, Jämtland, Sweden
(UTM 14297E 70066N).

Type stratum. – The uppermost 2 m of the Stein
Formation at the Andersön-B section, sample 99478
and 99479, *E. suecicus* Zone, Llanvirn.

Holotype. Repository PMO 165.319. A scandodontiform
element (Plate 9: 16).

Description. –

Drepanodontiform 1
The element is nongeniculate coniform with a plano-
convex base. The cusp is laterally compressed with a
keeled posterior margin. The basal margin is straight.
The basal cavity is relatively shallow with a triangular
outline in a lateral view. The apex is situated close to
the anterior margin. The antero-basal angle varies
commonly between 75 and 90° but occasionally may
be as low as 50°. The angle between the distal part of
the cusp and the upper margin is about 85°. The cusp
is reclined or recurved.

Drepanodontiform 2
The element is nongeniculate, laterally compressed
coniform with a slightly plano-convex base. Cusp has a
keeled posterior margin. The anterior margin of the
base is occasionally keeled. The basal margin is weakly
sinuous. The basal cavity is shallow with a triangular
outline in a lateral view. The apex is situated close to
the anterior margin. The antero-basal angle varies
between 90 and 100°. The angle between the distal
part of the cusp and the upper margin varies between
100 and 120°. The cusp is proclined.

Drepanodontiform 3
The element is nongeniculate, laterally compressed
coniform. The cusp is keeled both posteriorly and
anteriorly. The basal margin is straight. The basal
cavity is shallow with a triangular outline in a lateral
view. The apex is situated close to the anterior margin.
The antero-basal angle varies between 70 and 80°. The
angle between the distal part of the cusp and the
upper margin is 90 – 100°. The cusp is proclined or
suberect.

Scandodontiform
The element is a strongly inwardly bowed, geniculate
coniform, which is laterally compressed. Keels are
present both posteriorly and anteriorly. An inner-lateral
carina may be present in the anterior part of the cusp.
The basal cavity is shallow with the apex situated close
to the anterior margin. The antero-basal angle is about
70°, whereas the angle between the posterior margin of
the cusp and the upper margin is approximately 90°.
The cusp is suberect.

Oistodontiform
The element is geniculate, coniform. The cusp is
smooth and two or three times longer than the upper
margin of the base. Element is keeled. The angle
between the posterior margin of the cusp and the upper
margin of the base approximates 70°. The base develops
a characteristic triangular, antero-basal extension. The
cusp is reclined.

Remarks. – It was noted by Stouge (1984, p. 56) that
Paltodus? cf. *jemtlandicus* Löfgren (*sensu* Stouge) (=
drepanodontiform and oistodontiform elements of
Minimodus n. gen.) was dark brown with a black
coloured base. This is also the case in the specimens
described in this study. The plano-convex drepanodontiform 1 element of *Minimodus poulseni* n. sp.
resembles the drepanodontiform element of *Costiconus*
spp., but the latter is characterised by a clearly deeper
basal cavity, and a more irregular surface on the convex
side of the base. The oistodontiform element of
Triangulodus maocaopus Zhang (1998b) is very similar
to the oistodontiform element of *Minimodus poulseni*
described here, but no other *Triangulodus* elements
were recorded in this stratigraphical level (from the *M.
ozarkodella* Zone to the *E. suecicus* Zone) in the present
study. Similarly, no other *M. poulseni* elements were
reported by Zhang (1998b). Therefore, it is probable
that the oistodontiform elements belong to two
different evolutionary lineages, although they are
more or less identical.

Occurrence. – Andersön-B, Steinsodden, Røste. *M. ozarkodella* Zone – *E. suecicus* Zone.

Material. – 32 drepanodontiform 1, 26 drepanodontiform 2, 101 drepanodontiform 3, 5 scandodontiform and 72 oistodontiform elements.

Genus *Nordiora* n. gen.

Type species. – *Nordiora torpensis* n. sp.

Derivation of name. – After the Nordic countries (*Norden* in Danish) in combination with *ora* (Latin: margin, coast) in reference to its common occurrence in the distal parts of the Baltoscandian shelf.

Diagnosis. – *Nordiora* n. gen. is a septimembrate genus comprising angulate pectiniform Pa elements, pastinate pectiniform Pb elements, a geniculate, coniform M element and ramiform S elements including alate Sa elements, tertiopedate Sb elements, bipennate Sc elements and quadriramate Sd elements. P and S elements are denticulated, whereas the M element is adenticulate.

Remarks. – *Nordiora* includes *N. torpensis* n. sp., *Nordiora* n. sp. A and presumably *aff. Oepikodus? minutus* (McTavish) *sensu* Ethington & Clark (1982, pp. 62–63; Pl. 6:23, 24, 27, 28 [only]). In addition, the "early form" of *P. polonicus* (Dzik 1978) *sensu* Dzik (1994, Fig. 25: b; Pl. 17: 19-25) belongs to *Nordiora*.

Nordiora n. gen. is closely related to the Early Ordovician genus *Gothodus* Lindström 1955 *emended* Bagnoli & Stouge (1997), but *Gothodus* lack the laterally compressed, angulate pectiniform element, here interpreted as the Pa element. It is possible that *Nordiora* evolved from *Gothodus* during the late Arenig. *Nordiora* has also several characters in common with *Phragmodus* Branson & Mehl 1933 (see e.g. Bergström in Robison 1981, p. W129), but also important differences occur. *Nordiora* S elements are distinguished from those of *Phragmodus* by lack of the large denticle on the posterior process and the distinct bend just below it, and by having only weakly thickened processes.

Dzik (1978, Fig. 2) indicated that *Phragmodus polonicus*, possibly is the ancestor of the *Phragmodus* stock with a coniform M element, and that this species evolved from *Microzarkodina ozarkodella*. The present author agrees that *Phragmodus polonicus* is the most likely ancestor of the Middle Ordovician *Phragmodus* species. However, *P. polonicus* and *Nordiora* spp. are morphologically much closer to *Gothodus crassulus* (Lindström) emend. Bagnoli & Stouge (1997) than to *Microzarkodina ozarkodella* and probably evolved from this species.

Dzik (1978) chose a Pb ("ozarkodiniform") element from the Llanvirn *E. robustus* Zone as the holotype of *P. polonicus*, and noted that the species occurs from the *E. robustus* Zone to the *A. inaequalis* Zone. Later, Dzik (1994) distinguished between an early form (Dzik 1994, Pl. 18: 1-7; Fig. 25b) and a late form (Dzik 1994, Pl. 17: 19-25; Fig. 25c). The present author agrees that the "late form" including the holotype is a *Phragmodus* but believe that the "early form" is a *Nordiora*, although the angulate Pa element has not been observed so far.

Nordiora torpensis n. sp.

Pl. 10: 1–13

Derivation of name. – After the Torpa district in southern central Norway.

Type locality. – Road-cut at Røste, Øst-Torpa (UTM grid NN614583).

Type stratum. – 0.1 m above the base of the Stein Formation, sample 97759, *B. medius* – *H. holodentata* Zone, Lower Llanvirn.

Holotype. – Repository PMO 165.334. A Sd element (Pl. 10: 12).

Diagnosis. – A *Nordiora* species that is characterised by a Pa element which has up to five denticles on the lateral process. The Pb, Sa, Sb and Sd elements have distinct, narrow lateral costae on the cusp.

Description. –

Pa
Element is angulate pectiniform. The anterior process and the posterior process are of nearly equal length. The anterior process is almost straight and slightly bowed downward. Both processes bear four or five large, compressed, discrete denticles that are triangular in lateral view. The angle between the posterior and anterior process is 140°–150°. The cusp is erect or reclined.

Pb
The Pb element is pastinate pectiniform. The anterior, downward-directed process is adenticulate. The posterior process is denticulated. Denticles are compressed and discrete with a triangular outline in lateral view. The anterior process is broken in all specimens at hand. The cusp is erect.

M
The element is geniculate coniform and bears no denticles. The basal margin is slightly sinuous. Cusp is

reclined with an inner lateral carina. The angle between the upper margin of the base and the posterior margin of the cusp is about 60°.

Sa

Sa element is symmetrical, alate ramiform with denticles on the posterior process. The proximal denticles are erect, whereas distal are reclined. Lateral processes are bowed backward and adenticulate. Sharp, antero-lateral costae continue from the cusp onto the lateral processes. The angle between the lateral processes and the posterior process is about 22° in a lateral view. The cusp varies from proclined to recurved.

Sb

The Sb elements are asymmetrical, modified quadriramate ramiforms with a distinct, sharp outer costa. An indistinct inner costa is present in some specimens. The costa is curved downward and backward and continues into a short, adenticulate process. The short, adenticulate anterior process points downward. The denticulated posterior process is subhorisontal or bowed weakly downward. The denticles are weakly reclined, discrete and compressed with a triangular outline. The angle between the anterior and posterior process is 60–72°. The cusp varies from proclined to recurved.

Sc

The element is bipennate with a relatively long, denticulated posterior process and a short, downward-directed anterior process. The element is acostate, but an indistinct inner-lateral carina may occur on the cusp. The basal margin is straight or weakly convex below the cusp. Denticles are reclined. The angle between the anterior and posterior process is about 60°.

Sd

The Sd element is a subsymmetrical, quadriramate ramiforms with one distinct costa on each of the lateral surfaces of the cusp. The costae are curved downward and backward and continues into short, adenticulate processes. The short, adenticulate anterior process points downward. The denticulated posterior process is subhorisontal or bowed weakly downward. The denticles are weakly reclined, discrete and compressed with a triangular outline. Cusp varies from proclined to recurved. The angle between the anterior and posterior process is about 60°.

Remarks. – *N. torpensis* n. sp. is separated from the younger *Phragmodus polonicus* Dzik (see Dzik 1978, p. 63, Text-fig. 5 and 1991, Fig. 8c) by the angulate Pa element, the occurrence of recurved ramiform elements, the wider angle between the lateral processes and the basal margin in Sb and Sd elements, and the slightly longer anterior process in Sb and Sd elements.

N. torpensis is somewhat similar to *aff. Oepikodus? minutus* (McTavish) *sensu* Ethington & Clark (1982) with

respect to the Pb ("prioniodiform") and Sb ("unicostate oepikodiform") elements but the remaining element types are clearly different. The *N. torpensis* M element has a shorter posterior process and the Sd element bears reclined denticles distally instead of erect ones. Moreover, the Pb element has a relatively wider and shorter cusp than in aff. *Oepikodus? minutus* (McTavish) *sensu* Ethington & Clark (1982).

Occurrence. – Andersön-B, Steinsodden, Røste, Jøronlia, Hestekinn. *B. medius – H. holodentata* Zone – *P. graeai* Zone.

Material. – 9 Pa, 35 Pb, 1 ? P, 21 M, 7 Sa, 27 Sb, 9 Sc, 24 Sd.

Nordiora n. sp. A

Pl. 10: 14–18

Description. – The basal cavity is shallow in all elements.

Pa

The element is angulate pectiniform with denticulated anterior and posterior processes. The angle between

---

Plate XI

1–3: *Oelandodus elongatus* (Lindström).
1. P element, × 85. Sample 97800, Andersön-A. PMO 165.341.
2. P element, × 30. Sample 97800, Andersön-A. PMO 165.342.
3. Sb element, × 65. Sample 97800, Andersön-A. PMO 165.343.

4–8: *Oepikodus evae* (Lindström).
4. P element, × 140. Sample 69682, Herram. PMO 165.344.
5. M element, × 90. Sample 69687, Steinsodden. PMO 165.345.
6. M element, × 90. Sample 97805, Andersön-A. PMO 165.346.
7. S element, × 90. Sample 97805, Andersön-A. PMO 165.347.
8. S element, × 160. Sample 69682, Herram. PMO 165.348.

9: *Oistodus lanceolatus* Pander.
9. Sb element, × 90. Sample 97810, Andersön-A. PMO 165.349.

10: *Oistodus tablepointensis* Stouge.
10. S element, × 80. Sample 99464, Andersön-B. PMO 165.350.

11–14: *Paltodus deltifer* (Lindström).
11. Drepanodontiform element, × 60. Sample 97757, Røste. PMO 165.351.
12. Oistodontiform element, × 185. Sample 97757, Røste. PMO 165.352.
13. Acodontiform element, × 105. Sample 97757, Røste. PMO 165.353.
14. Drepanodontiform element, × 110. Sample 97757, Røste. PMO 165.354.

15–17: *Paltodus cf. deltifer* (Lindström).
15. Acodontiform element, × 70. Sample 97664, Grøslii. PMO 165.355.
16. Drepanodontiform element, × 50. Sample 97664, Grøslii. PMO 165.356.
17. Oistodontiform element, × 80. Sample 97664, Grøslii. PMO 165.357.

PLATE XI

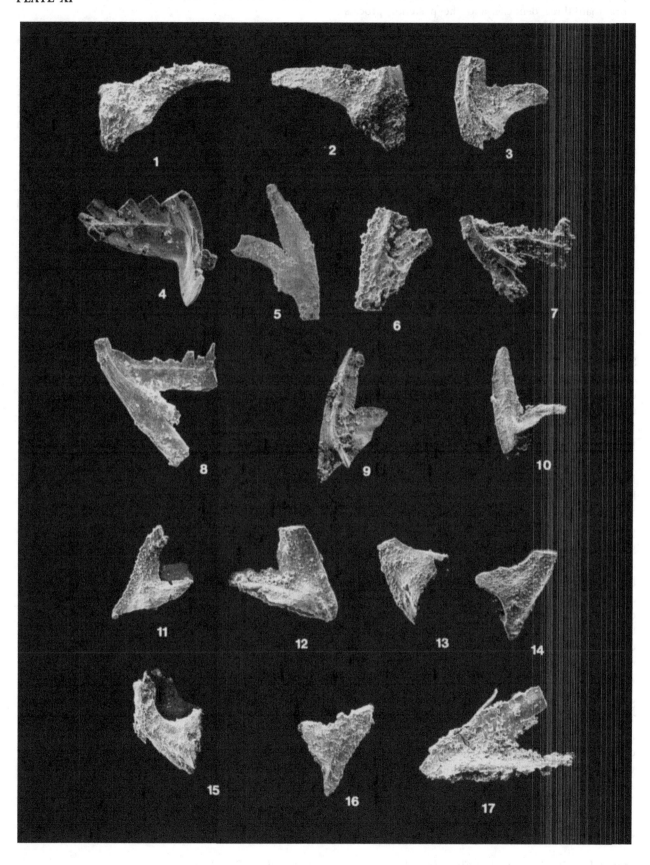

the processes is 100°. The anterior process carries more than three denticles and the posterior process more than two (both processes are broken). The cusp is slightly wider than the denticles. Both the cusp and denticles are laterally compressed. The denticles are fused proximally.

Pb

The element is pastinate pectiniform. The anterior, downward-directed process is adenticulate. The posterior process is denticulated. The denticles are compressed and discrete. The anterior and lateral processes are broken in the specimen at hand. The cusp is reclined.

M

The element is geniculate coniform and lacks denticles. The basal margin is slightly sinuous. Cusp is reclined with an indistinct inner lateral carina. The angle between the upper margin of the base and the posterior margin of the cusp is 80°.

Sa

Sa element is asymmetrical, alate ramiform with denticles on all processes. Every third or fourth denticle are more than twice as wide and long as the intervening denticles on the posterior process forming a "hindeodellid denticulation". The lateral processes are bowed backward. Sharp, postero-lateral costae continue from the cusp onto the lateral processes. The angle between the lateral and posterior processes is about 30° in a lateral view. Cusp is strongly recurved.

Sb

Sb elements are tertiopedate ramiforms with a distinct, sharp outer costa that continues downward into the lateral process. The adenticulate anterior process forms an angle of 75° with the posterior process. The denticulated posterior process is subhorisontal or bowed weakly downward. Denticles are reclined, discrete and compressed with a triangular outline making a "hindeodellid pattern". The cusp is recurved.

Sd

The Sd element is strongly asymmetrically quadriramate with four denticulated processes. The denticulated posterior process is subhorisontal or bowed weakly downward. Denticles on the posterior process are suberect and form a "hindeodellid denticulation". The cusp is recurved.

Occurrence. – Andersön-B. *M. ozarkodella* Zone.

Material. – 1 Pa, 2 Pb, 1 M, 3 Sb, 1 Sc, 2 Sd.

Genus *Oelandodus* van Wamel, 1974

Type species. – *Oistodus elongatus* Lindström, 1955

Remarks. – Van Wamel (1974, p. 71) interpreted the apparatus as trimembrate, comprising elongatiform, oistodontiform and triangulariform elements. These elements were assigned, respectively, to P, M and S elements by Bagnoli & Stouge (1988) and this notation tentatively has been used herein.

A slightly different apparatus reconstruction was introduced by Stouge & Bagnoli (1988). They interpreted the elongatiform and triangulariform element as S elements, the oistodontiform as a M element and added a new element type in the P position (Stouge & Bagnoli 1988, p. 120). It was not possible to recognise the P element *sensu* Stouge & Bagnoli within the present material. Therefore, the notation of Bagnoli & Stouge (1988) is used provisionally, awaiting a needed revision of *Oelandodus*.

Oelandodus elongatus (Lindström, 1955)

Pl. 11: 1 – 3

Synonymy. –

M
1955a *Oistodus elongatus* n. sp.–Lindström, p. 574; Pl. 4: 32, 33; Text-fig. 5b.

Multielement
1974 *Oelandodus elongatus* (Lindström) – van Wamel, pp. 71 – 72; Pl. 7: 1 – 4.
1988 *Oelandodus elongatus* (Lindström) – Bagnoli *et al.*, Pl. 39: 4, 8, 11.
1988 *Oelandodus elongatus* (Lindström) – Bergström, p. 11.
1993a *Oelandodus elongatus* (Lindström) – Löfgren, Fig. 9p.
1993b *Oelandodus elongatus* (Lindström) – Löfgren, Fig. 5P.
1994 *Oelandodus elongatus* (Lindström) – Löfgren, Fig. 8: 36, 37.
1995 *Oelandodus elongatus* (Lindström) – Lehnert, p. 98; Pl. 5: 25, 26.

Remarks. – The individual elements of *Oelandodus elongatus* were described by van Wamel (1974) and the apparatus was illustrated by van Wamel (1974) and Bagnoli & Stouge (1988).

Occurrence. – Andersön-A, -B. *P. proteus* Zone – *P. elegans* Zone.

Material. – 3 P, 3 M, 36 S.

Genus *Oepikodus* Lindström, 1955

Type species. – Oepikodus smithensis Lindström, 1955

Remarks. – The *Oepikodus* multielement apparatus was established by Lindström (1971), who included three element types in the apparatus. Recently, it was suggested by Stouge & Bagnoli (1988) (based on more than 5000 elements) that the apparatus is seximembrate, comprising Pa, Pb, M, Sb, Sc and Sd elements.

Oepikodus evae (Lindström, 1955)

Pl. 11: 4–8

Synonymy. –

P
1955a *Prioniodus evae* n. sp. – Lindström, pp. 589–590; Pl. 6: 4–10.

S
1955a *Oepikodus smithensis* n. sp. – Lindström, pp. 571–572; Pl. 6: 1–3.

M
1955a *Oistodus longiramis* n. sp. – Lindström, p. 579; Pl. 4: 35–37.

Multielement
1971   *Prioniodus evae* Lindström – Lindström, p. 52; Figs. 13–14.
1974   *Oepikodus evae* (Lindström) – van Wamel, p. 74; Pl. 6: 15–17.
1988   *Oepikodus evae* (Lindström) – Stouge & Bagnoli, p. 121; Pl. 5: 1–7 (*cum. syn.*).
1988   *Oepikodus evae* (Lindström) – Bergström, Pl. 1: 13–16.
1993b *Oepikodus evae* (Lindström) – Löfgren, Fig. 6A–C.
1994   *Oepikodus evae* (Lindström) – Löfgren, Fig. 8: 13–15.
? 1995a *Oepikodus evae* (Lindström) – Lehnert, pp. 99–100; Pl. 20B: 1.

Description. –

Pa and Pb, general remarks
Cusp and denticles are striated. The elements are characterised by a thickened upper part of the posterior process. The basal cavity is shallow. The cusp and denticles are carinate on both lateral surfaces.

Pa
The Pa element is pastinate characterised by an anterior process directed downward and one antero-lateral pro-

cess. The posterior process is straight or weakly twisted. The posterior process bears 5–8 denticles. The element is faintly asymmetrical seen from above in the anterior-posterior direction. The cusp is reclined or recurved.

Pb
The Pb element is distinguished from the Pa by a more laterally directed antero-lateral process, that makes an angle of about 90° with the cusp (Stouge & Bagnoli 1988). The cusp is erect or recurved.

Sb, Sc, Sd, general remarks
The S elements are ramiforms with a posterior process with hindeodelloid denticulation. The posterior process is bent weakly downward. The anterior process is adenticulate. The basal cavity is shallow through the entire process. The cusp is relatively long, laterally compressed and proclined or weakly almost erect. The S elements make a symmetry-transition series from elements bearing two indistinct lateral costae to elements characterised by two well-developed lateral costae. The denticles and cusp are striated.

Sb
The element is asymmetric "modified tertiopedate" with anterior and posterior processes. A distinct lateral costa, which downwardly forms a basal-lateral process-like extension, occurs on the outer surface of the element. A similar but more weakly developed costa is placed on the inner side of the element.

Sc
The Sc element is symmetrical bipennate. An indistinct costa is placed on each lateral surface of the cusp.

Sd
The element is "modified quadriramate" with one costa located on each lateral surface of the cusp. The costae are situated bilaterally symmetrical or weakly asymmetrical. The two costae make lateral, process-like extensions downwardly.

M
The M element is geniculate with a long, reclined cusp and a undenticulated, anterior extension of the base. The specimens agree closely with the description of Lindström (1955a).

Remarks. – The Sc element figured by Lehnert (1995; Pl. 20B:1) lacks the distinct lateral costae on the cusp seen in the Scandinavian specimens at hand.

Occurrence. – Andersön-A, Herram, Steinsodden. *O. evae* Zone – *M. flabellum – D. forceps* Zone. In addition, one (probably reworked) broken specimen was recorded from the base of the *P. zgierzensis* Zone at Steinsodden.

Material. – 36 Pa, 22 Pb, 15 M, 17 Sb, 7 Sc, 22 Sd.

Genus *Oistodus* Pander, 1856

Type species. – Oistodus lanceolatus Pander, 1856

Oistodus lanceolatus Pander, 1856

Pl. 11: 9

Synonymy. –

M

pt. 1856   *Oistodus lanceolatus* n. sp. – Pander, p. 27; Pl. 2:
                  18 (only), *non* 19 (= Gen. et sp. indet.).

1955a   *Oistodus lanceolatus* Pander – Lindström, p. 577;
             Pl. 3: 58 – 60.

1974   *Oistodus lanceolatus* Pander – Viira, Pl. 4: 2.

Sa

1955a *Oistodus delta* n. sp. – Lindström, p. 573; Pl. 3: 3–9.

1974   *Oistodus delta* Lindström – Viira, Pl. 4: 3.

Sb

pt. 1955a *Oistodus triangularis* n. sp. – Lindström, p. 581;
                 Pl. 4: 14 – 16 (only).

Sc

1856 *Oistodus lanceolatus* n. sp. – Pander, p. 27; Pl. 2: 17
        (only).

Sd

pt. 1955a *Oistodus triangularis* n. sp. – Lindström, p. 581;
                 Pl. 4: 17 – 18 (only).

1974   *Oistodus triangularis* Lindström – Viira, Pl. 2: 4.

Multielement

1971   *Oistodus lanceolatus* Pander – Lindström, p. 38.

1973   *Oistodus lanceolatus* Pander – Lindström in Zieg-
             ler, pp. 195 – 196, Oistodus – plate 1: 1 – 3.

1974   *Oistodus lanceolatus* Pander – van Wamel, p. 75;
             Pl. 1: 15 – 17.

1976   *Oistodus lanceolatus* Pander – Dzik, Fig. 21a, b.

1978   *Oistodus lanceolatus* Pander – Löfgren, pp. 63 –
             64; Pl. 1: 26 – 28.

1988   *Oistodus* aff. *lanceolatus* Pander – Bagnoli &
             Stouge, pp. 211 – 212; Pl. 40: 1 – 4.

1988   *Oistodus* aff. *lanceolatus* Pander – Stouge &
             Bagnoli, p. 123; Pl. 6: 1 – 8.

1988   *Oistodus lanceolatus* Pander – Bergström, Pl. 2:
             17 – 19.

1991   *Oistodus lanceolatus* Pander – Dzik, Fig. 10.

1993b *Oistodus lanceolatus* Pander – Löfgren, Fig. 6V – X.

1994   *Oistodus lanceolatus* Pander – Löfgren, Fig. 6:
             38 – 40.

1997   *Oistodus lanceolatus* Pander – Bagnoli & Stouge,
             pp. 148 – 150.

Remarks. – Lindström (1971, p. 38) reconstructed the

apparatus and distinguished three element types of *Oistodus lanceolatus* Pander. More recently, it was suggested
that the species was seximembrate (Sweet 1988, Fig. 5: 13;
Dzik 1991, Fig. 10). Most elements were described by
Lindström (1955a). It may be added that Sd elements are
distinguished from Sb elements by a more distinct, sharp
lateral process-like extension. The P element is similar to
the Sc element, but is characterised by a more straight cusp
and a less sinuous basal margin than the latter (Dzik 1991).
Dzik (1991, p. 295) named the two element types,
respectively, sp – oz (P) and ke – hi (Sc).

Oistodus aff. *lanceolatus* Pander *sensu* Bagnoli & Stouge
(1988, pp. 211 – 212) was regarded as different from
O. lanceolatus because of dissimilarities in the degree of
curvature between Panders original drawings of specimens from St. Petersburg, Russia and Bagnoli & Stouge's
material from Öland. Bergström (1988), however,

Plate XII

1–2: *Paltodus subaequalis* Pander.
1. Drepanodontiform element, × 105. Sample 99451, Andersön-B.
      PMO 165.358.
2. Oistodontiform element, × 100. Sample 99451, Andersön-B. PMO
      165.359.

3: *Paltodus* cf. *subaequalis* Pander.
3. Oistodontiform element, × 85. Sample 97757, Røste. PMO
      165.360.

4–5: *Panderodus sulcatus* (Fåhræus).
4. Falciform element, × 60. Sample 99479, Andersön-B. PMO
      165.361.
5. ? Arcuatiform element, × 80. Sample 99479, Andersön-B. PMO
      165.362.

6a: *Panderodus* sp. A.
6a. Arcuatiform element, × 65. Sample 97660, Glöte. PMO 165.363.

6b: *Panderodus* sp. B.
6b. Arcuatiform element, × 70. Sample 97660, Glöte. PMO 165.364.

7–10: *Paracordylodus gracilis* Lindström.
7. P element, × 155. Sample 97800, Andersön-A. PMO 165.365.
8. M element, × 95. Sample 97800, Andersön-A. PMO 165.366.
9. S element, × 65. Sample 97800, Andersön-A. PMO 165.367.
10. S element, × 45. Sample 97800, Andersön-A. PMO 165.368.

11: *Parapaltodus simplicissimus* Stouge.
11. Drepanodontiform element, × 55. Sample 99464, Andersön-B.
      PMO 165.369.

12: *Paroistodus horridus* (Barnes & Poplawski).
12. Drepanodontiform element, × 100. Sample 69676, Steinsodden.
      PMO 165.370.

13–17: *Paroistodus numarcuatus* (Lindström).
13. Drepanodontiform element, × 130. Sample 97757, RøstePMO
      165.371.
14. Oistodontiform element, × 110. Sample 97757, Røste. PMO
      165.372.
15. Drepanodontiform element, × 200. Sample 97668, Grøslii. PMO
      165.373.
16. Oistodontiform element, × 100. Sample 97668, Grøslii. PMO
      165.374.
17. Drepanodontiform element, × 155. Sample 97668, Grøslii. PMO
      165.375.

PLATE XII

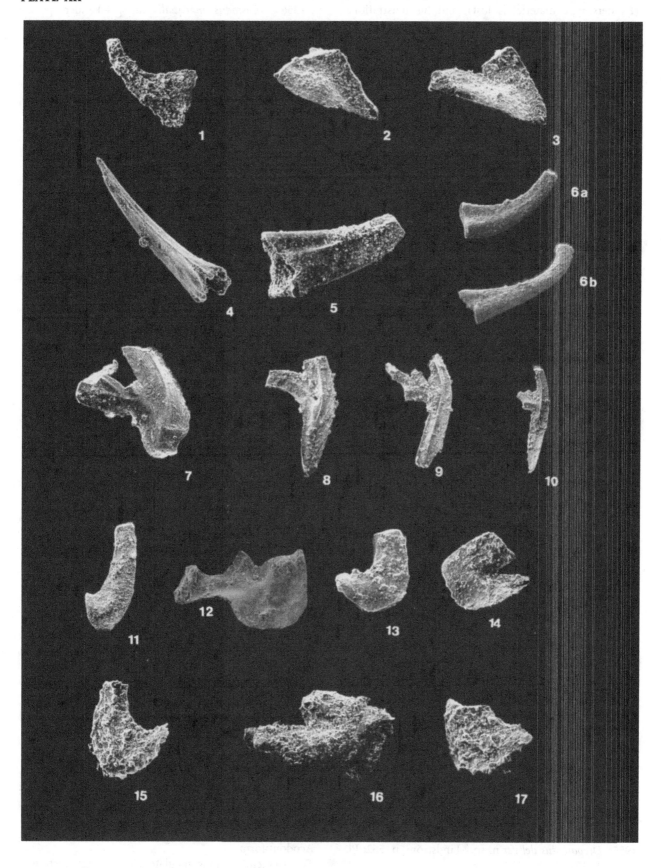

examined material from the St. Petersburg region, Russia (Panders type material is lost), and his illustrations (Bergström 1988, Pl. 2: 17–19) are similar to *O. lanceolatus* known elsewhere from Baltoscandia. This favours that Panders (1856) original illustrations, in particular that of the M element (Pander 1856, Pl. 2: 18) may be slightly distorted.

Occurrence. – Andersön-A, -B. *O. evae* Zone – basal part of the *P. rectus* – *M. parva* Zone.

Material. – 2 P, 1 Sa, 5 Sb, 1 Sc.

Oistodus tablepointensis Stouge, 1984

Pl. 11: 10

Synonymy. –

1984 *Oistodus? tablepointensis* n. sp. – Stouge, p. 75; Pl. 14: 3–12 (*cum. syn.*).

Remarks. – *Oistodus tablepointensis* was fully described by Stouge (1984).

Occurrence. – Andersön-B, Steinsodden. *P. graeai* Zone – *E. suecicus* Zone.

Material. – 4 S elements.

Genus *Paltodus* Pander, 1856

Type species. – *Paltodus subaequalis* Pander, 1856

Remarks. – Lindström (1971, 1977) included three element types in the *Paltodus* apparatus. The apparatus, however, presumably comprises one geniculate and three nongeniculate, coniform element types, including oistodontiform, drepanodontiform, acodontiform, and erectiform elements (Bagnoli & Stouge 1988). Löfgren (1997) reconstructed the apparatus as basically septimembrate consisting of 1 Sa, 2 Sb, 2 Sc, 2 Sd, 2 Pa, 2 Pb and 6 M elements (in total 17 elements).

Paltodus deltifer (Lindström, 1955)

Pl. 11: 11–14

Synonymy. –

Drepanodontiform
1955a *Drepanodus deltifer* n. sp. – Lindström, p. 562; Pl. 2: 42–43.

Oistodontiform
? 1856 *Oistodus inaequalis* n. sp. – Pander, p. 27; Pl. 2: 37.
pt. 1955a *Oistodus inaequalis* Pander – Lindström, p. 576; Pl. 3: 52, 55–56 (only).

Multielement
1988 *Paltodus deltifer* (Lindström) – Bagnoli & Stouge, p. 213; Pl. 40, 7–12 (*cum. syn.*).
1993a *Paltodus deltifer* (Lindström) – Löfgren, Fig. 9q, r.
1997b *Paltodus deltifer deltifer* (Lindström) – Löfgren, p. Fig. 5Z – AG.
1998 *Paltodus deltifer* (Lindström) – Bednarczyk, Pl. 1: 1, 21.

Remarks. – *Paltodus deltifer* was described by Lindström (1955a) and Bagnoli & Stouge (1988).

Occurrence. – Steinsodden, Røste. *P. deltifer* Zone – *P. proteus* Zone.

Material. – 3 drepanodontiform, 2 acodontiform, 3 oistodontiform.

Paltodus cf. *deltifer* (Lindström, 1955)

Pl. 11: 15–17

Synonymy. –

cf. 1955a *Drepanodus deltifer* n. sp. – Lindström, p. 562; Pl. 2: 42–43.
1989 *Paltodus deltifer deltifer* (Lindström) – Bruton, Harper & Repetski, p. 237; Fig. 5X – AD.
1995a "*Oistodus*" *inaequalis* (Lindström) – Lehnert, pp. 100–101; Pl. 1: 13.

Description. –

Drepanodontiform
The element is nongeniculate coniform with a smooth, suberect to slightly recurved cusp. The base is large and triangular in lateral view and forms an angle of 70–85° between the anterior and upper margin. An inwardly twisted flare may be developed on the anterior margin. The antero-basal corner is characterised by a small triangular extension that makes an angle of 30–40°.

Erectiform
The element is nongeniculate coniform with a proclined cusp. The posterior and anterior keels are sharp, whereas the low antero-lateral keel is more rounded. The basal margin is slightly convex.

Acodontiform
The element is nongeniculate coniform characterised

by a plano-convex base and a recurved cusp. An outer-lateral, median carina extends from the cusp to the straight basal margin. The angle between the anterior and upper margin is about 70°.

Oistodontiform

The element is geniculate coniform and inwardly flexed. The posterior margin of the cusp makes an angle of about 50° with the upper margin of the base. The antero-basal angle is about 30°. The cusp develops a median, rounded carina. The basal cavity is shallow.

Remarks. – The large base characterised by the wide angle between the upper and anterior margins of the drepanodontiform element distinguishes this species from *Paltodus deltifer* (Lindström) *sensu stricto*. The oistodontiform element figured by Lehnert (1995a, Pl. 1: 13) has a median, rounded carina that is typical of *Paltodus* cf. *deltifer*.

Occurrence. – Grøslii. *P. deltifer* Zone.

Material. – 7 drepanodontiform, 1 erectiform, 1 acodontiform and 4 oistodontiform elements.

Paltodus subaequalis Pander, 1856

Pl. 12: 1–2

Synonymy. –

Drepanodontiform
1856 *Paltodus subaequalis* n. sp. – Pander, p. 24; Pl. 1: 24; Text-fig. 4a.

Suberectiform
pt. 1955a *Distacodus peracutus* n. sp. – Lindström, pp. 555–556; Text-fig. 5d (only).
pt. 1987 *Distacodus peracutus* Lindström – Olgun, p. 48; Pl. 6J (only).

Oistodontiform
pt. 1955a *Oistodus inaequalis* Pander – Lindström, pp. 576–577; Pl. 3: 53, 54, 57 (only).

Multielement
1971 *Paltodus inconstans* Lindström – Lindström, p. 45; Text-fig. 8.
pt. 1974 *Drepanoistodus inconstans* (Lindström) – van Wamel, p. 67; Pl. 3: 11, 13–15 (only).
1977 *Paltodus subaequalis* Pander – Lindström, pp. 427–428, *Paltodus*- plate 1: 7–9.
1985 Paltodus subaequalis Pander – Löfgren, Fig. 4V – Y.

1988 *Paltodus subaequalis* Pander – Bagnoli & Stouge, pp. 214–215; Pl. 40: 13–17 (*cum. syn.*).
1988 *Paltodus subaequalis* Pander – Bergström, Pl. 2: 20–21.
1993a *Paltodus subaequalis* Pander – Löfgren, Fig. 9h–j.
1994 *Paltodus subaequalis* Pander – Löfgren, Fig. 6: 31–34.
1997b *Paltodus subaequalis* Pander – Löfgren, Fig. 5AO–AW.

Remarks. – The reconstruction of the *P. subaequalis* apparatus by van Wamel (1974) is followed here, but with the modification that the suberectiform element is excluded and transferred to *Drepanoistodus forceps* (Lindström). See van Wamel (1974) for description of the remaining elements.

Occurrence. – Andersön-A, -B. *P. proteus* Zone.

Material. – 3 drepanodontiform and 3 oistodontiform elements.

Paltodus cf. *subaequalis* Pander, 1856

Pl. 12: 3

Synonymy. –

cf. 1856 *Paltodus subaequalis* n. sp. – Pander, p. 24; Pl. 1: 24; Text-fig. 4a.
1985 *Paltodus* cf. *subaequalis* Pander – Löfgren, p. 127; Fig. 4R–U.
cf. 1988 *Paltodus subaequalis* Pander – Bagnoli & Stouge, pp. 214–215; Pl. 40: 13–17 (*cum. syn.*).
1997b *Paltodus* cf. *subaequalis* Pander – Löfgren, Fig. 5AH–AN.

Remarks. – The *P.* cf. *subaequalis* drepanodontiform elements lack the lateral costae which typifies *P. subaequalis*, whereas the oistodontiforms are similar to this species. *P.* cf. *subaequalis* was described by Löfgren (1985).

Occurrence. – Steinsodden, Røste. *P. deltifer* Zone – *P. proteus* Zone.

Material. – 6 drepanodontiform, 3 acodontiform and 4 oistodontiform elements.

Genus *Panderodus* Ethington, 1959

Type species. – *Paltodus unicostatus* Branson & Mehl, 1933.

Remarks. – The skeletal apparatus is seximembrate, comprising arcuatiform, graciliform, truncatiform, aequaliform, falciform, and tortiform elements (Sansom *et al.* 1994).

Panderodus sulcatus (Fåhræus, 1966)

Pl. 12: 4 – 5

Synonymy. –

Falciform
1966 *Paltodus sulcatus* n. sp. – Fåhræus, p. 25; Pl. 3: 9a – b.

Arcuatiform
1966 *Panderodus gracilis* (Branson & Mehl) – Fåhræus, p. 26; Pl. 3: 14a – b.

Multielement
1978 *Panderodus sulcatus* (Fåhræus) – Löfgren, p. 67; Pl. 8: 7 – 9.
1994 *Panderodus sulcatus* (Fåhræus) – Dzik, p. 59; Pl. 12: 21 – 28; Text-fig. 3a.
1998b *Panderodus sulcatus* (Fåhræus) – Zhang, pp. 78 – 79; Pl. 13: 1 – 5.

Remarks. – The present material leaves nothing to add to the descriptions of *Panderodus sulcatus* by Fåhræus (1966) and Löfgren (1978).

Occurrence. – Andersön-B. *E. suecicus* Zone.

Material. – 1 falciform and 2 arcuatiform elements.

Panderodus sp. A

Pl. 12: 6a

Description. –

General remarks
Elements are intensively striated near the basal margin. The basal cavity is deep. The cusp is proclined.

Graciliform
The element is slender, nongeniculate coniform. Two antero-lateral carinae extend from the basal margin to the upper part of the cusp resulting in a subsymmetrical outline.

Arcuatiform
The element is slender, nongeniculate coniform. An inner, antero-lateral costa extends from the basal margin to the upper part of the cusp. A median, lateral, rounded carina is situated on the outer surface of the element.

Tortiform?
Laterally compressed, nongeniculate element. The inner side is typified by a median, basal groove and antero-basal carina. The outer surface is smooth.

Falciform
Relatively wide, laterally compressed, nongeniculate coniform element. The anterior and posterior margins are rounded. A median, inner lateral groove extends from the basal margin to the cusp. The outer side is smooth.

Remarks. – Elements of *Panderodus* sp. A and *Panderodus* sp. B (see below) are distinguished from elements of *Panderodus sulcatus* by their deeper basal cavities.

Occurrence. – Glöte, *Pygodus anserinus* Zone.

Material. – 1 graciliform, 1 arcuatiform, 1 ?tortiform, 2 falciform and 2 indeterminable elements.

Panderodus sp. B

Pl. 12: 6b

Remarks. – *Panderodus* sp. B resembles *P.* sp. A but is separated from this by a deeper basal cavity, which extends almost to the tip of the cusp. The graciliform elements are longer and more slender than the corresponding elements in *Panderodus* sp. A. See also *Panderodus* sp. A.

Occurrence. – Glöte. *Pygodus anserinus* Zone.

Material. – 5 graciliform, 2 arcuatiform, 2 falciform.

Genus *Paracordylodus* Lindström, 1955

Type species. – *Paracordylodus gracilis* Lindström, 1955

Remarks. – *Paracordylodus* is trimembrate and comprises P ("cordylodontiform"), S ("paracordylodontiform"), and M ("oistodontiform") elements (van Wamel 1974; Stouge & Bagnoli 1988).

Paracordylodus gracilis Lindström, 1955

Pl. 12: 7–10

Synonymy. –

S

1955a *Paracordylodus gracilis* n. sp. – Lindström, p. 584; Pl. 6: 11–12.

M

1955a *Oistodus gracilis* n. sp. – Lindström, p. 576; Pl. 5: 1,2.

Multielement

1974 *Paracordylodus gracilis* Lindström – van Wamel, p. 77; Pl. 4: 11–13.

1988 *Paracordylodus gracilis* Lindström – Stouge & Bagnoli, p. 126; Pl. 8: 16–19 (*cum. syn.*).

1993b *Paracordylodus gracilis* Lindström – Löfgren, Fig. 5E–G.

1994 *Paracordylodus gracilis* Lindström – Löfgren, Fig. 8: 17–22.

1995a *Paracordylodus gracilis* Lindström – Lehnert, p. 105; Pl. 4: 3.

Remarks. – The specimens are identical with those described by Lindström (1955a, S and M elements) and van Wamel (1974, P element).

Occurrence. – Herram, Andersön-A, -B. *P. proteus* Zone – *M. flabellum* – *D. forceps* Zone.

Material. – 8 P, 61 M, 206 S.

Genus *Parapaltodus* Stouge, 1984

Type species. – *Parapaltodus simplicissimus* Stouge, 1984

Remarks. – The apparatus comprises drepanodontiform and scandodontiform elements (Stouge 1984).

Parapaltodus simplicissimus Stouge, 1984

Pl. 12: 11

Synonymy. –

1984 *Parapaltodus simplicissimus* n. sp. – Stouge, p. 48; Pl. 1: 20–21, 26–28 (*cum. syn.*).

1990 *Parapaltodus simplicissimus* Stouge – Stouge & Bagnoli, p. 21; Pl. 7: 13–14.

1995a *Parapaltodus simplicissimus* Stouge – Lehnert, p. 105; Pl. 7: 1, Pl. 11: 13.

Remarks. – The species was described in detail by Stouge (1984).

Occurrence. – Andersön-B. *B. medius* - *H. holodentata* Zone.

Material. – 1 drepanodontiform element.

Genus *Parapanderodus* Stouge, 1984

Type species. – *Parapanderodus arcuatus* Stouge, 1984

Remarks. – *Parapanderodus* was interpreted as trimembrate by Stouge & Bagnoli (1990), who distinguished between symmetrical, asymmetrical and slender elements (coniform A, B and C elements, respectively, in this study).

Parapanderodus quietus Bagnoli & Stouge, 1997

Not figured

Synonymy. –

Coniform A

1985 *Scolopodus gracilis* (Ethington & Clark) – Löfgren, Fig 4AI (only).

Coniform B

? 1985 *Scolopodus gracilis* (Ethington & Clark) – Löfgren, Fig 4AH (only).

Coniform C

1978 *Scolopodus* aff. *gracilis* (Ethington & Clark) – Löfgren, p. 110; Pl. 8: 10A–B.

1985 *Scolopodus gracilis* (Ethington & Clark) – Löfgren, Fig 4AG, AJ, ?AK (only).

Multielement

1990 *Parapanderodus* n. sp. A – Stouge & Bagnoli, p. 21; Pl. 7: 1–4.

1997 *Parapanderodus quietus* n. sp. – Bagnoli & Stouge, pp. 150–151; Pl. 5: 6–10.

Remarks. – The specimens found agree with Stouge & Bagnoli's (1990) description.

Occurrence. – Glöte. ? *B. medius* - *H. holodentata* Zone.

Material. – 1 coniform A and 1 coniform C element.

Genus *Paroistodus* Lindström, 1971

Type species. – *Oistodus parallelus* Pander, 1856

Remarks. – The apparatus is basically bimembrate comprising nongeniculate (drepanodontiform) and geniculate (oistodontiform) elements (Lindström 1971). The nongeniculate elements make a curvature transition series (Dzik 1976; Fåhræus & Hunter 1985b), and was considered as septimembrate by Löfgren (1997a), who distinguished six different types of the drepanodontiform element (as Pa, Pb, Sa, Sb, Sc and Sd) and one oistodontiform element (as M).

Paroistodus horridus (Barnes & Poplawski, 1973)

Pl. 12: 12

Synonymy. –

Drepanodontiform

1973 *Cordylodus horridus* n. sp. – Barnes & Poplawski, pp. 771 – 772; Pl. 2: 16 – 18.

1984 *Cordylodus? horridus* Barnes & Poplawski – Stouge, p. 45; Pl. 1: 1 – 11 (*cum. syn.*).

1984 "*Cordylodus*" *horridus* Barnes & Poplawski – Nowlan & Thurlow, p. 291; Pl. 1: 4, 7 – 8.

1991 "*Cordylodus*" *horridus* Barnes & Poplawski – McCracken, p. 49; Pl. 1: 1.

1992 *Cordylodus horridus* Barnes & Poplawski – Nicoll, Fig. 4: 3 – 4.

1995a *Paroistodus horridus* Barnes & Poplawski – Lehnert, p. 108; Pl. 8: 15 – 16, Pl. 11: 2, 12; Pl. 13: 7.

Oistodontiform

1969 *Oistodus venustus* (Stauffer) – Bradshaw, p. 1150; Pl. 134: 4 – 7.

1984 *Paroistodus? cf. originalis* (Sergeeva) – Stouge, p. 56; Pl. 5: 1 – 4 (*cum. syn.*).

1995a *Paroistodus cf. originalis* (Sergeeva) – Lehnert, p. 110; Pl. 11: 6.

Multielement

pt. 1998b *Paroistodus horridus* (Barnes & Poplawski) – Zhang, pp. 79 – 80; Pl. 13: 13 – 21 (*non* Pl. 13: 11 – 12, which are believed to belong to *Paroistodus originalis*).

Remarks. – The nongeniculate *Paroistodus horridus* element was fully described by Stouge (1984), who distinguished between three different element types. He tentatively referred the taxon to *Cordylodus? horridus* but was of the opinion that it presumably belongs to another genus. The oistodontiform element conforms with the description of Stouge (1984, p. 56). The oistodontiform element differs from *Paroistodus originalis* by having a concave basal margin instead of a convex or straight. *P. horridus* differs from the other

Paroistodus species by the presence of denticles on the nongeniculate elements. The Cow Head slope deposits of Newfoundland contain drepanodontiform elements transitional between *Paroistodus originalis* and *P. horridus* (Stouge pers. comm. 1988), and similar intermediate elements have also been documented from Argentina (Albanesi & Barnes 1996). The elements have not yet developed denticles, but the white matter forms denticle-like structures close to the upper margin. Accordingly, it seems verified that the form-species *Cordylodus horridus* belongs to *Paroistodus* and not to *Cordylodus*. Löfgren (1995a) observed denticles on the upper margin of *Paroistodus* already in the *O. evae* Zone, but apparently the development was not successful for the genus before the Early Llanvirn.

Occurrence. – Andersön-A, -B, Steinsodden, Røste. *B. medius - H. holodentata* Zone – *E. suecicus* Zone.

Material. – 11 drepanodontiform, 1 oistodontiform.

Paroistodus numarcuatus (Lindström, 1955)

Pl. 12: 13 – 17

Plate XIII

1–3: *Paroistodus proteus* (Lindström).
1. Drepanodontiform element, × 100. Sample 97800, Andersön-A. PMO 165.376.
2. Oistodontiform element, × 105. Sample 97800, Andersön-A. PMO 165.377.
3. Drepanodontiform element, × 80. Sample 97800, Andersön-A. PMO 165.378.

4–5: *Paroistodus parallelus* (Pander).
4. Oistodontiform element, × 60. Sample 97800, Andersön-A. PMO 165.379.
5. Drepanodontiform element, × 100. Sample 97668, Grøslii. PMO 165.380.

6–7: *Paroistodus originalis* (Sergeeva).
6. Drepanodontiform element, × 55. Sample 99458, Andersön-B. PMO 165.381.
7. Oistodontiform element, × 75. Sample 99458, Andersön-B. PMO 165.382.

8–11: *Periodon aculeatus* Hadding.
8. Pa element, × 90. Sample 99484, Andersön-B. PMO 165.383.
9. M element, × 65. Sample 99484, Andersön-B. PMO 165.384.
10. Sb element, × 125. Sample 99484, Andersön-B. PMO 165.385.
11. Sd element, × 100. Sample 99484, Andersön-B. PMO 165.386.

12–16: *Periodon flabellum* (Lindström).
12. Pa element, × 155. Sample 69687, Steinsodden. PMO 165.387.
13. M element, × 155. Sample 69685, Herram. PMO 165.388.
14. Sb element, × 85. Sample 69687, Steinsodden. PMO 165.389.
15. Sc element, × 170. Sample 69685, Herram. PMO 165.390.
16. Sd element, × 100. Sample 97685, Herram. PMO 165.391.

PLATE XIII

Synonymy. –

Drepanodontiform

1955a *Drepanodus numarcuatus* n. sp. – Lindström, pp. 564 – 565; Pl. 2: 48 – 49; Text-fig. 3I.

1955a *Drepanodus amoenus* n. sp. – Lindström, p. 558; Pl. 2: 25(?) – 26; Text-fig. 4.

1974 *Drepanodus numarcuatus* Lindström – Viira, p. 69; Text-figs. 15, 70 – 71.

1974 *Drepanodus amoenus* Lindström – Viira, p. 66; Text-figs. 15, 65.

1974 Transitional morphotypes between *Drepanodus numarcuatus* Lindström and *Drepanodus amoenus* Lindström – Viira, Text-fig. 15.

Oistodontiform

pt. 1955a *Oistodus parallelus* Pander – Lindström, pp. 579 – 580; Pl. 4: 27 – 29; Text-fig. 3N (only).

Multielement

1971 *Paroistodus numarcuatus* (Lindström) – Lindström, p. 46; Text-fig. 8.

pt. 1974

Drepanoistodus numarcuatus (Lindström) – van Wamel, pp. 67 – 69; Pl. 3: 8 (only).

1974 *Paroistodus amoenus* (Lindström) – van Wamel, pp. 78 – 79; Pl. 7: 8 – 11.

1981 *Paroistodus numarcuatus* (Lindström) – Lindström (*in* Ziegler), pp. 227 – 228, *Paroistodus*-plate 2: 7, 8.

1988 *Paroistodus numarcuatus* (Lindström) – Bagnoli & Stouge (*in* Bagnoli *et al.*), Pl. 39: 13 – 15.

1988 *Paroistodus numarcuatus* (Lindström) – Stouge & Bagnoli, pp. 127 – 128; Pl. 8: 8 – 11.

1989 *Paroistodus numarcuatus* (Lindström) – Bruton, Harper & Repetski, Fig. 5O – P, ?Q.

1993a *Paroistodus numarcuatus* (Lindström) – Löfgren, Fig. 8e – h.

1994 *Paroistodus numarcuatus* (Lindström) – Löfgren, Fig. 6: 1 – 3.

1995a *Paroistodus numarcuatus* (Lindström) – Lehnert, pp. 108 – 109; Pl. 1: 14.

1997a *Paroistodus numarcuatus* (Lindström) – Löfgren, pp. 921 – 922; Text-fig. 2O – U, 3A – G, 4A – K.

1997b *Paroistodus numarcuatus* (Lindström) – Löfgren, Fig. 4B.

1998 *Paroistodus numarcuatus* (Lindström) – Bednarczyk, Pl. 1: 2.

Remarks. – The observed specimens are small and poorly preserved. The elements conform with the descriptions of Lindström (1955a; *in* Ziegler 1981); Stouge & Bagnoli (1988) and Löfgren (1997a). The species differs from the other *Paroistodus* species for example by the lack of inverted basal cavity (Lindström *in* Ziegler 1981) and by a relatively smaller base in drepanodontiform elements.

Occurrence. – Steinsodden, Røste, Grøslii. *P. deltifer* Zone – *P. proteus* Zone.

Material. – 14 drepanodontiform and 9 oistodontiform elements.

Paroistodus originalis (Sergeeva, 1963)

Pl. 13: 6 – 7

Synonymy. –

Drepanodontiform

1960 *Distacodus* sp. – Lindström, Fig. 3: 6.

Oistodontiform

1963 *Oistodus originalis* n. sp. – Sergeeva, p. 98; Pl. 7: 8, 9; Text-fig. 4.

Multielement

1971 *Paroistodus originalis* (Sergeeva) – Lindström, p. 48; Text-figs. 8, 12.

pt. 1987 *Drepanoistodus basiovalis* (Sergeeva) – Olgun, p. 49; Pl. 6: GH (only).

1990 *Paroistodus originalis* (Sergeeva) – Stouge & Bagnoli, p. 22; Pl. 7: 5 – 10 (*cum. syn.*).

1991 *Paroistodus originalis* (Sergeeva) – Rasmussen, pp. 279 – 280; Fig. 7L – M.

1994 *Paroistodus originalis* (Sergeeva) – Löfgren, Fig. 6: 13 – 16.

1995a *Paroistodus originalis* (Sergeeva) – Lehnert, pp. 109 – 110; Pl. 7: 2 – 3, Pl. 13: 5.

1995a *Drepanoistodus basiovalis* (Sergeeva) – Lehnert, pp. 83 – 84; Pl. 9: 9, Pl. 10: 5, 12.

? 1995a *Drepanoistodus forceps* (Lindström) – Lehnert, p. 84; Pl. 13: 2 (only).

1995b *Paroistodus originalis* (Sergeeva) – Löfgren, Fig. 7u – y.

1997a *Paroistodus originalis* (Sergeeva) – Löfgren, pp. 926 – 927; Pl. 1: 13 – 16, 18 – 20, 22 – 33; Text-fig. 5H – O.

pt. 1998b *Paroistodus originalis* (Sergeeva) – Zhang, pp. 79 – 80; Pl. 13: 11 – 12 (only).

Remarks. – The specimens correspond with the descriptions of Löfgren (1978, 1997a) and Lindström (*in* Ziegler 1981). *Paroistodus originalis* resembles *P. proteus*, but *P. originalis* rarely includes carinate elements (Stouge & Bagnoli 1988).

Occurrence. – Andersön-A, -B, Herram, Steinsodden, Røste, Jøronlia. (? *O. evae* Zone) *M. flabellum - D. forceps* Zone – *P. graeai* Zone.

Material. – 813 drepanodontiform and 566 oistodontiform elements.

Paroistodus parallelus (Pander, 1856)

Pl. 13: 4 – 5

Synonymy. –

Oistodontiform
1856 *Oistodus parallelus* n. sp. – Pander, p. 27; Pl. 2: 40.

Drepanodontiform
1941 *Acodus expansus* n. sp. – Graves & Ellison, p. 8; Pl. 1: 6.

Multielement
1971 *Paroistodus parallelus* (Pander) – Lindström, p. 47; Figs. 8, 11.
1978 *Paroistodus parallelus* (Pander) – Löfgren, pp. 68 – 69; Pl. 1: 18 – 21.
1988 *Paroistodus parallelus* (Pander) – Bagnoli and Stouge (*in* Bagnoli *et al.*), p. 128; Pl. 8: 2, 6, 7 (*cum. syn.*).
1988 *Paroistodus parallelus* (Pander) – Bergström, Pl. 2: 22 – 23.
1991 *Paroistodus parallelus* (Pander) – Rasmussen, p. 281; Fig. 7N.
1993b *Paroistodus parallelus* (Pander) – Löfgren, Fig. 5H – K.
1997a *Paroistodus parallelus* (Pander) – Löfgren, pp. 923 – 924; Pl. 1: 1 – 12, 17, 21; Text-fig. 5A – G.
1998 *Paroistodus parallelus* (Pander) – Bednarczyk, Pl. 1: 5, 24.
1998 *Paroistodus originalis* (Sergeeva) – Bednarczyk, Pl. 1: 18, 6?

Remarks. – *Paroistodus parallelus* was described by Löfgren (1978, 1997a) and Stouge & Bagnoli (1988). It differs from the other *Paroistodus* species by having a distinct, longitudinal carina or costa on each of the lateral surfaces of the drepanodontiform element. The specimens referred to *Paroistodus originalis* by Bednarczyk (1998) include a drepanodontiform element with a median costa on the side of the figured element (Bednarczyk 1998, Pl. 1:18), which is typical of *Paroistodus parallelus*.

Occurrence. – Andersön-A, -B. *P. proteus* Zone – *O. evae* Zone.

Material. – 30 drepanodontiform and 4 oistodontiform elements.

Paroistodus proteus (Lindström, 1955)

Pl. 13: 1 – 3

Synonymy. –

Drepanodontiform
1955a *Drepanodus proteus* n. sp. – Lindström, pp. 566 – 567; Pl. 3: 18 – 21; Text-fig. 2a – f, j.
1960 *Drepanodus proteus* Lindström – Lindström, Fig. 1: II: 5.

Oistodontiform
1960 *Oistodus parallelus* Pander – Lindström, Fig. 1: II: 6.
1974 *Oistodus parallelus* Pander – Viira, Pl. 2: 27.

Multielement
1971 *Paroistodus proteus* (Lindström) – Lindström, pp. 46 – 47; Text-figs. 8, 10.
1988 *Paroistodus proteus* (Lindström) – Stouge & Bagnoli, pp. 128 – 129; Pl. 8: 1, 3 – 5 (*cum. syn.*).
1988 *Paroistodus proteus* (Lindström) – Bergström, Pl. 2: 24.
1993a *Paroistodus proteus* (Lindström) – Löfgren, Fig. 9m – n.
1993b *Paroistodus proteus* (Lindström) – Löfgren, Fig. 5L – O.
1994 *Paroistodus proteus* (Lindström) – Löfgren, Fig. 6: 4 – 9.
1995a *Paroistodus proteus* (Lindström) – Lehnert, p. 110; Pl. 3: 9 – 10.
1997a *Paroistodus proteus* (Lindström) – Löfgren, pp. 922 – 923; Text-figs. 3H – N, 4L – AB.

Remarks. – *Paroistodus proteus* was described and discussed by Lindström (1955a), Stouge & Bagnoli (1988) and Löfgren (1997a).

P. proteus is distinguished from *P. parallelus* by the lack of lateral costae or carinae in the drepanodontiform elements. It differs from *P. numarcuatus* by the common presence of inverted basal cavity (Lindström 1971, *in* Ziegler 1981). The oistodontiform element of *P. proteus* is indistinguishable from that of *P. parallelus*. The two species occur together in the Tøyen Formation at Andersön, and in this case the oistodontiform elements have been divided proportionally between the two species in relation to the respective number of drepanodontiform elements.

Occurrence. – Andersön-A, -B, Steinsodden. *P. proteus* Zone – *P. elegans* Zone.

Material. – 78 drepanodontiform and 30 oistodontiform elements.

Genus *Periodon* Hadding, 1913

Type species. – *Periodon aculeatus* Hadding, 1913

Remarks. – Bergström & Sweet (1966) reconstructed the multielement genus *Periodon* Hadding and included six distinct element-types in the apparatus, but already Lindström (1964) showed that the four ramiform elements form a symmetry-transition series. Recently, it was shown that *Periodon* is septimembrate (Stouge 1984). Subsequent studies by Stouge & Bagnoli (1988) confirmed that it comprises seven element types (Pa, Pb, M, Sa, Sb, Sc, Sd). The present paper follows the elemental interpretation of Stouge & Bagnoli (1988) with the exception that the "loxognathiform" element is placed in the Sb position instead of the Sd position. This is because it is morphologically closer to the alate Sa element than the "periodontiform" (Sd herein) element. It has been documented that the evolutionary early *Periodon* species produced fewer denticles than the late species (Serpagli 1974; van Wamel 1974; Löfgren 1978; Stouge 1984; Stouge & Bagnoli 1988). In addition to analysis of other characters, this has revealed that the following five denticulated Arenig to upper Llanvirn *Periodon* species may be recognised in Baltoscandia: *Periodon primus* Stouge & Bagnoli 1988, *P. selenopsis* (Serpagli 1974), *P. flabellum* (Lindström 1955), *P. zgierzensis* Dzik 1976, *P. macrodentata* (Graves & Ellison 1941) and *P. aculeatus* Hadding 1913. *Periodon primus* has not been observed in the present study.

Periodon aculeatus Hadding, 1913

Pl. 13: 8 – 11

Synonymy. –

Pa

1962 *Prioniodina macrodentata* (Graves & Ellison) – Sweet & Bergström, p. 1240; Pl. 171: 7, 8.

1964 *Prioniodina macrodentata* (Graves & Ellison) – Hamar, p. 278; Pl. 3: 28; Text-fig. 4: 19.

1966 *Prioniodina pulcherrima* Lindström – Webers, p. 60; Pl. 12: 9, 11.

Pb

1962 *Ligonodina tortilis* n. sp. – Sweet & Bergström, pp. 1230 – 1231; Pl. 170: 13, 14.

M

1962 *Falodus prodentatus* (Graves & Ellison) – Sweet & Bergström, pp. 1227 – 1229; Pl. 170: 2, 3; Text-fig. 2B.

1964 *Falodus prodentatus* (Graves & Ellison) – Hamar, p. 265; Pl. 4: 9, 10; Text-fig. 4: 18.

pt. 1964 *Oistodus venustus* Stauffer – Hamar, p. 269; Pl. 3: 9 (only).

1966 *Falodus prodentatus* (Graves & Ellison) – Webers, p. 56; Pl. 12: ?6, 7.

Sb

1964 *Loxognathus grandis* Ethington – Hamar, Pl. 3: 22, 24, 26, 27; Text-fig. 4: 15.

Sc

? pt. 1962 *Periodon aculeatus* Hadding – Sweet & Bergström, p. 1235; Pl. 171: 3 (only).

1964 *Paracordylodus bergstroemi* n. sp. – Hamar, p.273; Pl. 3: 20, 23, 25; Text-fig. 6: 12.

Sd

1913 *Periodon aculeatus* n. sp. – Hadding, p. 33; Pl. 1: 14.

1955b *Periodon aculeatus* Hadding – Lindström, p. 10; Pl. 22: 10 – 11, 14 – 16, 35.

pt. 1962 *Periodon aculeatus* Hadding – Sweet & Bergström, p. 1235; Pl. 171: 9 (only).

1964 *Periodon aculeatus* Hadding – Hamar, p. 274; Pl. 3: 17, 21.

1966 *Periodon aculeatus* Hadding – Webers, Pl. 12: 16.

Multielement

1974 *Periodon aculeatus* Hadding – Bergström, Riva & Kay, Pl. 1: 4 – 6.

? 1976 *Periodon aculeatus aculeatus* Hadding – Dzik, p. 435; Text-fig. 34l – r.

1978 *Periodon aculeatus* Hadding – Bergström, Pl. 79: 3 – 5.

pt. 1978 *Periodon aculeatus* Hadding – Löfgren, pp. 74 – 75; Pl. 11: 19 – 26 (only).

1981 *Periodon aculeatus* Hadding – Nowlan, p. 12; Pl. 2: 7 – 10, Pl. 4: 1 – 9.

Plate XIV

1–8: *Periodon macrodentata* (Graves & Ellison).
1. Pa element, ×95. Sample 69675, Steinsodden. PMO 165.392.
2. Pa element, ×65. Sample 69675, Steinsodden. PMO 165.393.
3. Pb element, ×130. Sample 69675, Steinsodden. PMO 165.394.
4. M element, ×80. Sample 69675, Steinsodden. PMO 165.395.
5. Sa element, ×100. Sample 69675, Steinsodden. PMO 165.396.
6. Sb element, ×75. Sample 69675, Steinsodden. PMO 165.397.
7. Sc element, ×65. Sample 69675, Steinsodden. PMO 165.398.
8. Sd element, ×90. Sample 69675, Steinsodden. PMO 165.399.

9–11: *Periodon selenopsis* (Serpagli).
9. Pa element, ×90. Sample 97805, Andersön-A. PMO 165.400.
10. M element, ×80. Sample 97805, Andersön-A. PMO 165.401.
11. Sc element, ×70. Sample 97805, Andersön-A. PMO 165.402.

12–16: *Periodon zgierzensis* Dzik.
12. Pa element, ×95. Sample 69622, Steinsodden. PMO 165.403.
13. Pa element, ×95. Sample 69622, Steinsodden. PMO 165.404.
14. M element, ×135. Sample 69622, Steinsodden. PMO 165.405.
15. M element, ×65. Sample 69622, Steinsodden. PMO 165.406.
16. Sb element, ×110. Sample 69622, Steinsodden. PMO 165.407.

PLATE XIV

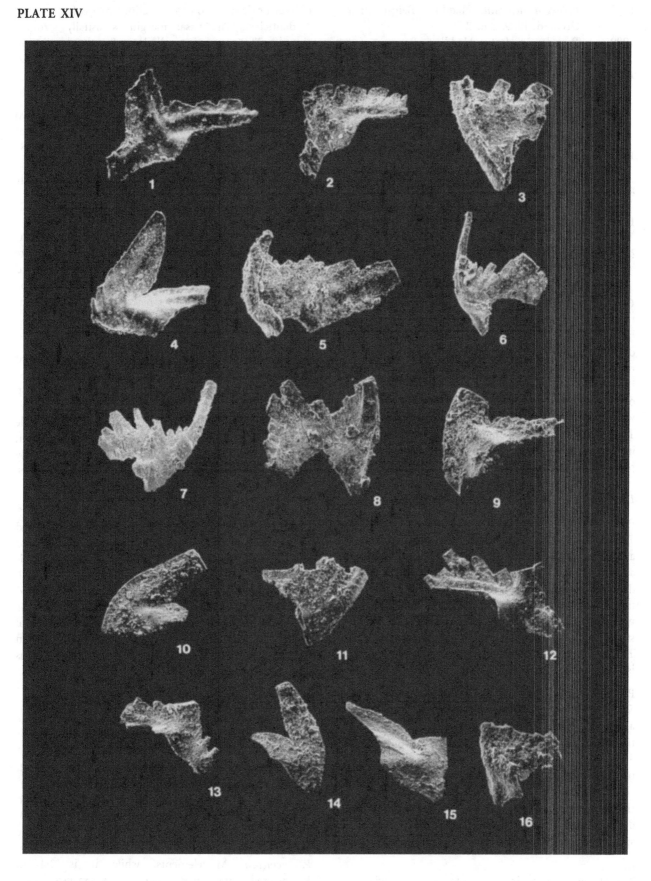

1985   *Periodon aculeatus* Hadding – Bergström & Orchard, Pl. 2.2: 6, 7.

1989   *Periodon aculeatus* Hadding – Rasmussen & Stouge, Fig. 3G – K.

1990   *Periodon aculeatus* Hadding – Bergström, Pl. 1: 15, 16.

1991   *Periodon aculeatus* Hadding – McCracken, p. 50; Pl. 2: 24 – 27, 31, 34, 35 (with additional synonymy).

1994   *Periodon aculeatus* Hadding – Dzik, p. 111; Pl. 24: 10 – 13; Text-fig. 31b.

pt. 1995a *Periodon aculeatus* Hadding – Lehnert, pp. 110 – 112; Pl. 16: 8, 9, 11 – 13 (only).

1997   *Periodon aculeatus* Hadding – Armstrong, pp. 774 – 775; Pl. 2: 13 – 21; Text-fig. 3.

? 1998b *Periodon aculeatus* Hadding – Zhang, pp. 80 – 81; Pl. 14: 1 – 8.

Comments to the synonymy list. – The elements referred to *Periodon aculeatus* by Löfgren (1978, Pl. 11:12 – 18), have both fewer denticles on the M elements and fewer denticles between the cusp and the largest denticle on the posterior process than *P. aculeatus* in the present interpretation. In addition, the ratio between the length of the basal margin and length of the cusp is about 3:4, which is typical for *Periodon macrodentata* (see below). The elements figured by Zhang (1998b, pl. 14: 1 – 8) are in general less advanced than *P. aculeatus* and do probably belong to *P. macrodentata*.

Description. –

Pa

The element is angulate pectiniform with denticulated anterior and posterior process. The inwardly-bowed anterior process commonly carries between 4 and 6 denticles. The posterior process commonly has five or more denticles. The angle between the two processes is typically 130 – 150°, but may decrease to about 100° in stratigraphically young specimens (*P. anserinus* Zone). The cusp is erect or slightly recurved. The basal margin is sinuous.

Pb

The Pb element is digyrate with a strongly twisted anterior process. The anterior process is slightly thickened at the lateral surfaces and usually has 3 or 4 denticles. The number of denticles on the posterior process is commonly more than 6. The cusp is usually erect, but may be weakly reclined. The basal margin is slightly sinuous. The element conforms with the description of the form-species *Ligonodina tortilis* n. sp. by Sweet & Bergström (1962, pp. 1230 – 1231).

M

The M element is geniculate coniform, usually with 3 –

6 anterior denticles on the anterior process (on average 4 denticles). The basal margin is usually strongly sinuous. The cusp is inwardly bent. See the description of *Falodus prodentatus* by Sweet & Bergström (1962, pp. 1227 – 1229) for further details.

Sa

The Sa element is alate with one posterior and two lateral processes, all of which are denticulated. It is not possible to estimate the number of denticles as the Sa elements are broken. The basal cavity is sometimes inverted. The cusp is erect, weakly proclined or weakly reclined.

Sb

The element is tertiopedate with one posterior, one anterior and one outer-lateral process. All processes are denticulated. The outer-lateral process bents downward and slightly backward. The basal cavity is inverted. The posterior process is usually twisted inward. The cusp is weakly recurved.

Sc

Sc element is dolabrate with a multidenticulated posterior process and an anterior process bearing 0 – 2 denticles. The number of denticles between the cusp and the biggest denticle vary usually between 6 and 8. The posterior process is often twisted. The angle between the antero-basal margin and the lower margin of the posterior process is 40 – 60°. The cusp is weakly reclined or recurved. The basal cavity is inverted. See Hamar (1964, p. 273) for further description.

Sd

The Sd element is modified tertiopedate characterised by a multidenticulated, twisted, posterior process and an anterior process that in some specimens is denticulated in the basal part. A lateral, small process-like extension forms the downward directed continuation of the outer costa or carina. The cusp is suberect or weakly reclined or recurved. The basal cavity is inverted. The element was described in further detail by Lindström (1955b, pp. 110 – 111) as *Periodon aculeatus* Hadding *sensu formo*.

Remarks. – The *Periodon aculeatus* Pa element differs from *Periodon macrodentata* (Graves & Ellison) by the relatively longer and laterally more thickened anterior process and a smaller basal sheath.

The Pb element of *P. aculeatus* is distinguished from the older species by having an angle of about 90° between the anterior and the posterior process, which is slightly less than observed by Stouge (1984, p. 83: 90 – 100°). In addition, Stouge (1984) noted that the length of basal margin:length of cusp ratio is 1:1 in *P. aculeatus* M elements, while it is 3:4 in *P. macrodentata* (= *P. aculeatus zgierzensis* sensu

Stouge 1984). These observations are confirmed in the present study.

Occurrence. – Høyberget, Engerdal, Andersön-B, -C, Steinsodden, Glöte. *E. suecicus* Zone – *P. anserinus-A. inaequalis* Subzone.

Material. – 46 Pa, 12 Pb, 58 M, 7 Sa, 10 Sb, 46 Sc, 29 Sd.

Periodon flabellum (Lindström, 1955)

Pl. 13: 12 – 16

Synonymy. –

Pa

1955a   *Prioniodina inflata* n. sp. – Lindström, p. 588; Pl. 6: ?26, 27.

1964   *Oulodus inflatus* (Lindström) – Lindström, p. 85; Fig. 30B ("prioniodina element").

pt. 1967 *Prioniodina inflata* Lindström – Viira, Text-fig. 1: 4 (only).

1974   *Prioniodina inflata* Lindström – Viira, Pl. 4: 18 – 20.

Pb

1964   *Oulodus inflatus* (Lindström) – Lindström, p. 85; Text-fig. 30A ("oulodus element").

? pt. 1967 *Prioniodina inflata* Lindström – Viira, Text-fig. 1: 3 (only).

Sa

1955a *Trichonodella flabellum* n. sp. – Lindström, p. 599; Pl. 6: 28 – 30.

1964   *Periodon flabellum* (Lindström) – Lindström, p. 83; Text-fig. 28D ("roundya element").

1967   *Trichonodella flabellum* Lindström – Viira, Text-fig. 1: 1.

1974   *Trichonodella flabellum* Lindström – Viira, Pl. 4: 8 – 9.

Sb

1964 *Periodon flabellum* (Lindström) – Lindström, p. 83; Fig. 28B ("cladognathodus element").

Sc

pt. 1955a *Prioniodina? deflexa* n. sp. – Lindström, p. 586; Pl. 6: 32 – 35 (only).

1960   *Prioniodina? deflexa* Lindström – Lindström, Fig. 2: 5.

1964   *Periodon flabellum* Lindström – Lindström, p. 83; Fig. 28A ("cordylodus element").

? pt. 1967 *Prioniodina inflata* Lindström – Viira, Text-fig. 1: 3 (only).

pt. 1974 *Prioniodina? deflexa* Lindström – Viira, Pl. 4: 10 (only).

Sd

pt. 1955a *Prioniodina? deflexa* n. sp. – Lindström, p. 586; Pl. 6: 31 (only).

1964   *Periodon flabellum* (Lindström) – Lindström, p. 83; Text-fig. 28C ("ligonodina element").

1967   *Prioniodina? deflexa* Lindström – Viira, Text-fig. 1: 2.

? pt. 1974 *Prioniodina? deflexa* Lindström – Viira, Pl. 4: 15, 16 (only).

M

1955a *Oistodus selene* n. sp. – Lindström, p. 508; Pl. 4: 19 – 20.

Multielement

1971   *Periodon flabellum* (Lindström) – Lindström, p. 57; Text-fig. 18.

? pt. 1974 *Periodon flabellum* (Lindström) – Serpagli, pp. 63 – 64; Pl. 14: 1 – 2, 4 – 7; Pl. 25: 7 – 8, 10 – 12.

pt. 1974 *Periodon flabellum* (Lindström) – van Wamel, pp. 80 – 81; Pl. 4: 19 (only).

pt. 1976 *Periodon flabellum* (Lindström) – Dzik, Fig. 34a – d (only).

pt. 1978 *Periodon flabellum* (Lindström) – Löfgren, p. 72; Pl. 11: 1 – 5 (only).

? 1978 *Periodon flabellum* (Lindström) – Fåhræus & Nowlan, p. 462; Pl. 3: 2 – 6; Text-fig. 5A – F.

1988   *Periodon flabellum* (Lindström) – Stouge & Bagnoli, p. 129; Pl. 9: 12 – 18.

1993b *Periodon flabellum* (Lindström) – Löfgren, Fig. 6M – P.

1994   *Periodon flabellum* (Lindström) – Löfgren, Fig. 8: 39 – 42.

? 1995b *Periodon flabellum* (Lindström) – Löfgren, Fig. 7: p – t.

1997   *Periodon* cf. *flabellum* (Lindström) – Bagnoli & Stouge, p. 151; Pl. 6: 1 – 7.

1997   *Periodon* sp. A – Bagnoli & Stouge, pp. 151 – 153; Pl. 6: 8, 12 – 18.

1997   *Periodon* sp. B – Bagnoli & Stouge, pp. 152; Pl. 8: 9 – 12.

pt. 1998 *Periodon flabellum* (Lindström) – Bednarczyk, Pl. 1: 7 (only).

Comments to the synonymy list. – The apparatus illustrated by Serpagli (1974) comprises Pa elements with three denticles in front of the cusp, M elements with maximum three denticles and S elements with three to four denticles between the cusp and the largest denticle at the posterior process. This indicates that it probably belongs to *Periodon zgierzensis* Dzik. The elements pictured by Fåhræus & Nowlan (1978) are more advanced than the *P. flabellum* at hand and are questionably referred to this species, but it is also possible that it belongs to *P. zgierzensis*. The elements included in *P. flabellum* by van Wamel (1974, Pl. 4: 14–18, 20) belong to *Periodon primus* (figs. 14, 18) and *P. selenopsis* (figs. 15–17, 20) (Stouge & Bagnoli 1988). *P. flabellum sensu* Löfgren (1978, Pl. 6–11 [*partim*]) belongs to *Periodon zgierzensis* Dzik. The Pa element figured by Löfgren (1995b, Fig. 7r) is here interpreted as a Pb element because of the strongly twisted anterior denticles. The M elements (Löfgren 1995b, Fig. 7s–t) seem to carry 2–3 weakly developed anterior denticles, which indicate that the figured specimens may belong to *Periodon zgierzensis* instead of *P. flabellum*. The interpretation of *P. flabellum* presented here is broader than that of Bagnoli & Stouge (1997) because the different variations described as *Periodon* cf. *flabellum*, *Periodon* sp. A and *Periodon* sp. B by Bagnoli & Stouge commonly occur in the same samples as the *P. flabellum sensu stricto* in the sections investigated here. The Sc element showed by Bednarczyk (1998, Pl. 1: 4 (only)) is more advanced than typical *P. flabellum* and does probably belong to *Periodon aculeatus* or *P. macrodentata*.

Remarks. – The material conforms with the descriptions of Lindström (1955a) and Stouge & Bagnoli (1988). It may be added that the Pa elements usually carry 1 (0–2) anterior denticles on the basal part of the anterior margin whereas the M element carries 0 or occasionally up to 2 denticles. In comparison, *P. zgierzensis* develops a denticulated anterior, "true" process on the Pa element. The Pa element figured in Pl. 13: 12 belongs to an advanced form of *P. flabellum*, which carries a tiny denticle-like extension in front of the anterior denticle. *P. flabellum* differs from *P. selenopsis* by having anterior denticles on the Pa element.

Occurrence. – Andersön-A, -B, Herram. *M. flabellum - D. forceps* Zone.

Material. – 21 Pa, 28 Pb, 42 M, 15 Sa, 17 Sb, 50 Sc, 18 Sd.

Periodon macrodentata (Graves & Ellison, 1941)

Pl. 14: 1–8

Synonymy. –

Pa

1941   *Ozarkodina   macrodentata   n.   sp.* – Graves & Ellison, p. 14; Pl. 2: 33, 35, 36.

1969   *Prioniodina   macrodentata* (Graves & Ellison) – Bradshaw, p. 1160; Pl. 137: 19.

1970   *Prioniodina   macrodentata* (Graves & Ellison) – Uyeno & Barnes, p. 113; Pl. 23: 12, 16.

Pb

1969   *Ligonodina   tortilis   (Sweet   &   Bergström)* – Bradshaw, pp. 1152–1153; Pl. 137: 18.

M

1941   *Oistodus prodentatus* n. sp. – Graves & Ellison, pp. 13–14; Pl. 2: 6, 22, 23, 28.

1969   *Falodus   prodentatus* (Graves   &   Ellison) – Bradshaw, p. 1151; Pl. 135: 16, 17.

1970   *Falodus prodentatus* (Graves & Ellison) – Uyeno & Barnes, p. 108; Pl. 22: 8, 14, 18.

1970   *Falodus* sp. – Uyeno & Barnes, p. 108; Pl. 22: 15.

Sb

1941   *Loxognathus   flabellata   n.   sp.* – Graves & Ellison, p. 12; Pl. 2: 29, 31, 32.

pt. 1969 *Periodon   aculeatus*   Hadding – Bradshaw, pp. 1159–1160; Pl. 137: 4, 5 (only).

pt. 1970 *Periodon aculeatus* Hadding – Uyeno & Barnes, p. 112; Pl. 23: 3, 5 (only).

Sd

pt. 1969 *Periodon   aculeatus*   Hadding – Bradshaw, pp. 1159–1160; Pl. 137: 1–3, 6 (only).

Plate XV

1–4, 7–8, 12: *Polonodus?* cf. *tablepointensis* Stouge.
1. Pa-1 element, × 50. Sample 99473, Andersön-B. PMO 165.408.
2. Pa-2 element, × 65. Sample 99473, Andersön-B. PMO 165.409.
3. Pb element, × 100. Sample 99473, Andersön-B. PMO 165.410.
4. Pb element, × 100. Sample 99473, Andersön-B. PMO 165.411.
7. Sb element, × 100. Sample 99473, Andersön-B. PMO 165.412.
8. Sb element, × 115. Sample 99473, Andersön-B. PMO 165.413.
12. Sb element, × 100. Sample 99479, Andersön-B. PMO 165.414.

5–6: *Polonodus clivosus* (Viira).
5. Pa-1 element, × 30. Sample 99465, Andersön-B. PMO 165.415.
6. M element, × 90. Sample 99465, Andersön-B. PMO 165.416.

9–10: *Polonodus* cf. *clivosus* (Viira).
9. Pa-1 element, × 55. Sample 99473, Andersön-B. PMO 165.417.
10. Pa-2 element, × 55. Sample 99473, Andersön-B. PMO 165.418.

11: *Polonodus* sp. A.
11. Pa-1 element, × 30. Sample 97730, Jøronlia. PMO 165.419.

13–16: *Prioniodus elegans* Pander.
13. Pa element, × 120. Sample 97804, Andersön-A. PMO 165.420.
14. Pb element, × 120. Sample 97804, Andersön-A. PMO 165.421.
15. M element, × 80. Sample 97804, Andersön-A. PMO 165.422.
16. Sb element, × 120. Sample 97804, Andersön-A. PMO 165.423.

17–19: *Protopanderodus graeai* (Hamar).
17. Scandodontiform element, × 70. Sample 69662, Steinsodden. PMO 165.424.
18. Symmetrical acontiodontiform element, × 75. Sample 69662, Steinsodden. PMO 165.425.
19. Symmetrical acontiodontiform element, × 75. Sample 69662, Steinsodden. PMO 165.426.

20–21: *Protopanderodus calceatus* Bagnoli & Stouge.
20. Scandodontiform element, × 75. Sample 99471, Andersön-B. PMO 165.427.
21. Acontiodontiform element, × 50. Sample 99471, Andersön-B. PMO 165.428.

PLATE XV

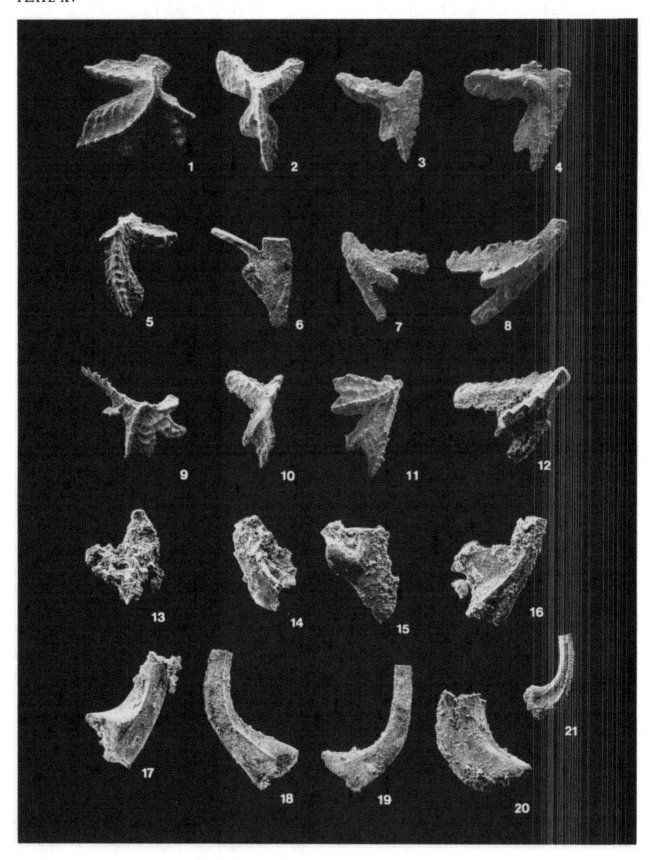

pt. 1970 *Periodon aculeatus* Hadding – Uyeno & Barnes, p. 112; Pl. 23: 1, 2?, 4, 6, 7 (only).

1970 *Periodon flabellum* (Lindström) – Uyeno & Barnes, p. 112; Pl. 23: 10?, 15.

Multielement

1984 *Periodon aculeatus zgierzensis* Dzik – Stouge, p. 82; Pl. 16: 1 – 15 (*cum. syn.*).

1985 *Periodon aculeatus* Hadding – Bergström & Orchard, p. 56; Pl. 2.2: 6,7.

pt. 1987 *Periodon aculeatus* Hadding – Pohler *et al.*, Fig. 21.10D – F (only).

pt. 1991 *Periodon aculeatus* Hadding – McCracken, p. 50; Pl. 1: 13, 20, 22, 25 – 28 (only).

pt. 1995a *Periodon aculeatus* Hadding – Lehnert, pp. 110 – 112; Pl. 10: 2; Pl. 11: 10, 11; Pl. 13: 9 – 12, ?Pl. 7: 9 (only).

? 1998b *Periodon aculeatus* (Hadding) – Zhang, pp. 80 – 81; Pl. 14: 1 – 8.

Remarks. – *Periodon macrodentata* was fully described by Stouge (1984) who referred it to *Periodon aculeatus zgierzensis* Dzik 1976. *Periodon macrodentata* is distinguished from *P. zgierzensis* by the distinctly thickened lateral surfaces in the Pa element, a more sinuous basal margin in the M element, and by having an angle between the anterior and posterior process in the Pb element at 90 – 100° instead of 100 – 130°. Moreover, *Periodon macrodentata* forms a true anterior process on the M element and has more denticles on the anterior margin of Pa and M elements than the stratigraphically older species (on average 3.75 and 3.40 respectively). *Periodon macrodentata* has been named after the form species *Ozarkodina macrodentata* Graves & Ellison 1941 instead of *Loxognathus flabellata* Graves & Ellison 1941 to avoid confusion with *Periodon flabellum* (Lindström 1955). The specimens figured by Zhang (1998b) as *P. aculeatus* are more similar to *P. macrodentata* (especially the Pb and M elements) and do probably belong to this species.

Occurrence. – Andersön-A, -B, -C, Steinsodden, Røste, Jøronlia, Haugnes, Hestekinn, Skogstad. *B. norrlandicus - D. stougei* Zone (upper part) – *E. suecicus* Zone.

Material. – 287 Pa, 162 Pb, 383 M, 82 Sa, 162 Sb, 281 Sc, 142 Sd.

Periodon selenopsis (Serpagli, 1974)

Pl. 14: 9 – 11

Synonymy. –

1974 *"Oistodus" selenopsis* n. sp. – Serpagli, pp. 56 – 57; Pl. 13: 4a – 5d; Pl. 23: 8, 9.

pt. 1974 *Periodon flabellum* (Lindström) – Serpagli, pp. 63 – 64; Pl. 14: 3a – b; Pl. 25: 9 (only).

1988 *Periodon selenopsis* (Serpagli) – Stouge & Bagnoli, pp. 130 – 131; Pl. 9: 5 – 11 (*cum. syn.*).

Remarks. – The apparatus of *Periodon selenopsis* was reconstructed and described by Stouge & Bagnoli (1988).

Occurrence. – Andersön-A, -B. *O. evae* Zone – basal *M. flabellum - D. forceps* Zone.

Material. – 3 Pa, 2 Pb, 12 M, 1 Sa, 5 Sb, 9 Sc.

Periodon zgierzensis Dzik, 1976

Pl. 14: 12 – 16

Original diagnosis. – "Trichonodelliform element with reduced denticulation on the lateral branches (as in *P. aculeatus aculeatus*), oistodontiform element without or with 2 denticles at the anterior margin at the most (as in *P. flabellum*)" (Dzik 1976).

Emended diagnosis. – *Periodon zgierzensis* Dzik is characterised by a Pa element which forms a laterally compressed, anterior process that carries two or more denticles. The M element is usually adenticulate but may carry up to three antero-basal denticles.

Synonymy. –

Multielement

1974 *Periodon flabellum* (Lindström) – Serpagli, p. 63; Pl. 14: 1 – 7, Pl. 25: 7 – 12.

1976 *Periodon aculeatus zgierzensis* n. ssp. – Dzik, p. 424; Pl. 44: 5, 6; Text-fig. 34E – K.

pt. 1978 *Periodon flabellum* (Lindström) – Löfgren, p. 72; Pl. 11: 6 – 11 (only).

? 1978 *Periodon flabellum* (Lindström) – Fåhræus & Nowlan, p. 462; Pl. 3: 3 – 6, ?2; Text-fig. 5A – C, F (only).

cf. 1978 *Periodon aculeatus* Hadding – Fåhræus & Nowlan, p. 462; Pl. 3: 1, 7 – 10, 13; Text-fig. 5G – L (only).

pt. 1978 *Periodon aculeatus - Periodon flabellum*, oistodiforms – Fåhræus & Nowlan, Pl. 3: 1 (only).

1980 *Periodon flabellum* (Lindström) – Merrill, Fig. 6: 36 – 39.

pt. 1987 *Periodon aculeatus* Hadding – Pohler *et al.*, Fig. 21.10A – C (only).

1990 *Periodon* sp. – Stouge & Bagnoli, p. 22; Pl. 7: 11 – 12.

1991   *Periodon aculeatus* ssp. A – Rasmussen, p. 281;
       Fig. 7Q, R.
1995a *Periodon flabellum* (Lindström) – Lehnert,
       p. 112; Pl. 7: 8, 10 – 11.
1995a *Periodon* sp. A – Lehnert, pp. 112 – 113; Pl. 7:
       6.
? 1995b *Periodon flabellum* (Lindström) – Löfgren,
       Fig. 7p – t.
1997   *Periodon* sp. C – Bagnoli & Stouge, pp. 152 –
       154; Pl. 6: 19 – 20.

Comments to the synonymy list. – The stratigraphically younger specimens referred to *Periodon flabellum* by Löfgren (1978, Pl. 11: 6 – 11) include a Pa element distinguished by a denticulated anterior process, which is characteristic of *P. zgierzensis*.

P. flabellum sensu Fåhræus & Nowlan (1978) have more anterior denticles in the M element than typical for that species and is here referred to *P. zgierzensis*.

The specimens placed in *Periodon aculeatus* by Fåhræus & Nowlan (1978) include a Pb element, which is less angulate than that of *P. aculeatus*. Additionally, the Pa element lacks the thickening of the anterior process that typifies this species. The specimens figured by Lehnert (1995a, Pl. 7: 8, 10, 11) display a more advanced denticulation than typical for *P. flabellum*, and should instead be referred to *P. zgierzensis*. The specimens figured by Löfgren (1995b, Fig. 7p – t) belong possibly to *P. zgierzensis*, but a precise species identification is difficult because of the lack of figured Pa elements (see *P. flabellum* for further comments).

Description. –

Pa
The element is bipennate or angulate with denticulated anterior and posterior processes. The inwardly-bowed anterior process commonly bears two or three denticles, but may occasionally have four (average = 2.74 in the Stein section). The posterior process typically have four or five denticles. The cusp is erect or slightly recurved. The basal margin is sinuous.

Pb
The Pb element is digyrate with a strongly twisted anterior process. The anterior process is slightly thickened at the lateral surfaces and has 3 or four denticles. The number of denticles on the posterior process varies commonly between four and six. The cusp is often usually erect, but may be weakly proclined or reclined. The basal margin is slightly sinuous.

M
The M element is geniculate coniform with 0 – 2, rarely 3 denticles in front of the cusp. The basal margin is weakly sinuous. The cusp is inwardly bent.

Sa
The Sa element is alate with one posterior and two lateral processes, all of which are denticulated. It is not possible to estimate the number of denticles as all Sa elements are broken. The basal cavity may be inverted. The cusp is erect, weakly proclined or weakly reclined.

Sb
The element is tertiopedate with one posterior, one anterior and one outer-lateral process. All processes are denticulated. The outer-lateral process bents downward and slightly backward, but it is not directed straight backward as typical in younger *Periodon* species. The basal cavity in mature specimens is distinctly inverted. The posterior process is usually twisted. The cusp is weakly recurved.

Sc
The Sc element is dolabrate with a multidenticulated posterior process commonly carrying about 4 or 5 denticles between the cusp and the largest denticle on the posterior process. The anterior margin bears 0 – 2 denticles. The posterior process is often twisted. Some specimens have an outer carina on the cusp like in *Periodon macrodentata* (= *Periodon aculeatus zgierzensis* Dzik *sensu* Stouge 1984). The cusp is weakly reclined or recurved.

Sd
The element is modified tertiopedate characterised by one multidenticulated, twisted, posterior process and one adenticulate anterior process. A small process-like extension, forms the downward directed continuation of the outer costa or carina. The cusp is weakly reclined or recurved.

Remarks. – The holotype of *Periodon zgierzensis* was described from an erratic boulder from the early Llanvirn *E. pseudoplanus* Zone of Dzik (1976) and represents the most advanced variation of the species as it is interpreted here. *Periodon zgierzensis* differs from *P. flabellum* by having a "true" anterior process on the Pa element, and from *P. macrodentata* by the common lack of denticles on the M element and the more laterally compressed processes on the Pa element. The more advanced *P. zgierzensis* specimens (see below) resemble *P. macrodentata* but may be distinguished from this by the more obtuse angle between the anterior and posterior processes in the Pb element.

Discussion. – Some of the stratigraphically older specimens of *P. zgierzensis* in the Andersön-A and Andersön-B sections (sample 97822 and 99458) comprise M and Pa elements with more denticles than the younger *P. zgierzensis* specimens and also include Pa elements with a thickened anterior process.

This tendency opposes the general trend within *Periodon* (Löfgren 1978). It is probable that the number of denticles also was influenced by environmental changes. Future studies will possibly show that the advanced *Periodon zgierzensis* morphotype constitutes a separate (eco?-)subspecies or species but because the amount of material showing this variation is very small, it has been tentatively included in *Periodon zgierzensis*.

Occurrence. – Andersön-A, -B, Herram, Steinsodden. *M. flabellum - D. forceps* Zone – *B. medius - H. holodentata* Zone (basal part).

Material. – 62 Pa, 27 Pb, 67 M, 13 Sa, 22 Sb, 60 Sc, 30 Sd.

Genus *Polonodus* Dzik, 1976

Type species. – *Ambalodus clivosus* Viira, 1974

Remarks. – The apparatus is septimembrate comprising planate Pa-1, Pa-2 and Pb, geniculate M, alate Sa, tertiopedate Sb, and quadriramate Sd elements. The apparatus was reconstructed and described by Löfgren (1990) who recognised six element types. The interpretation presented here agrees with that of Löfgren (1990) except that it includes a previously undescribed Pb element to the apparatus.

When Dzik (1976) erected the genus *Polonodus*, he selected the form-species *Ambalodus clivosus* Viira 1974 as type species and referred his illustrated specimens (Dzik 1976, Pl. 43: 1; Textfigs. 7, 29c, d) to this species (*P. clivosus*). However, the present author agrees with Zhang (1998b) that these four-lobed specimens are Pa-2 elements of the *Polonodus tablepointensis* group, and lack the distinct crests and ridges orientated perpendicular to the platform ledge, as is typical of *P. clivosus* (see Viira 1967, Fig. 3: 24a, b; Viira 1974: 37, 38, Pl. 8: 1; and Rasmussen 1991, Figs. 8A – C). Stouge (1984) noted that *Polonodus* comprises two main morphological groups, one including *P. clivosus* (Viira) *sensu* Stouge and *P. newfoundlandensis* and another comprising *P. tablepointensis*. Subsequently, it was suggested by Löfgren (1990, p. 254) that *P. clivosus sensu stricto* and *P. tablepointensis* Stouge may be separated from *P. newfoundlandensis* and *P. clivosus sensu* Löfgren (1978) (= *Polonodus* cf. *clivosus* herein) by lack of a blade-like process. The material at hand verifies this. Zhang (1998b) went one step further and erected the new genus *Dzikodus* for the *P. tablepointensis* group, which was distinguished from *Polonodus* by having "unpaired, markedly dissimilar pastiplanate elements" instead of "paired, mutually quite similar pastiniplanate ele-

ments". The present author agrees with Zhang (1998b) that it is likely that the genus *Polonodus* may be subdivided in more genera, and *Dzikodus* Zhang 1998 may well turn out to be a valid genus. However, because the apparatus of the *Polonodus* type species, *P. clivosus* (Viira), is still very poorly known, it has been chosen to maintain the name *Polonodus* for both morphological groups until the type species has been investigated in greater detail.

Polonodus clivosus (Viira, 1974)

Pl. 15: 5 – 6

Synonymy. –

1967	*Ambalodus* n. sp. – Viira, Fig. 3: 24a,b.
1974	*Ambalodus clivosus* n. sp. – Viira, p. 51; Pl. 8: 1; Text-figs. 37 – 38.
non 1974	*Ambalodus*? n. sp. – Viira, p. 52; Pl. 8: 2 – 3; Text-fig. 39.
non 1976	*Polonodus clivosus* (Viira) – Dzik, p. 432; Pl. 43: 1; Fig. 7, Fig. 29c – d.
non 1984	*Polonodus*? *clivosus* (Viira) – Stouge, p. 73; Pl. 16: 6 – 13.

Plate XVI

1–4: *Protopanderodus rectus* (Lindström).
1. Scandodontiform element, × 70. Sample 99458, Andersön-B. PMO 165.429.
2. Symmetrical acontiodontiform element, × 100. Sample 99458, Andersön-B. PMO 165.430.
3. Asymmetrical acontiodontiform element, × 100. Sample 99458, Andersön-B. PMO 165.431.
4. Asymmetrical acontiodontiform element, × 100. Sample 99458, Andersön-B. PMO 165.432.

5–8: *Protopanderodus robustus* (Hadding).
5. Scandodontiform element, × 50. Sample 99479, Andersön-B. PMO 165.433.
6. Symmetrical acontiodontiform element, × 75. Sample 99479, Andersön-B. PMO 165.434.
7. Symmetrical acontiodontiform element, × 40. Sample 99479, Andersön-B. PMO 165.435.
8. Asymmetrical acontiodontiform element, × 95. Sample 99479, Andersön-B. PMO 165.436.

9–10: *Protoprioniodus* sp. A.
9. P element, × 195. Sample 69682, Herram. PMO 165.437.
10: S element, × 235. Sample 69682, Herram. PMO 165.438.

11–12: *Protoprioniodus* cf. *costatus* (van Wamel).
11. P element, × 105. Sample 97799, Andersön-A. PMO 165.439.
12. S element, × 60. Sample 97799, Andersön-A. PMO 165.440.

13–17: *Pygodus anserinus* Lamont & Lindström.
13. Pa element, × 50. Sample 99488, Andersön-B. PMO 165.441.
14. Pb element, × 80. Sample 99488, Andersön-B. PMO 165.442.
15. Sa element, × 100. Sample 99488, Andersön-B. PMO 165.443.
16: Sb element, × 70. Sample 99488, Andersön-B. PMO 165.444.
17: Sd element, × 125. Sample 99488, Andersön-B. PMO 165.445.

PLATE XVI

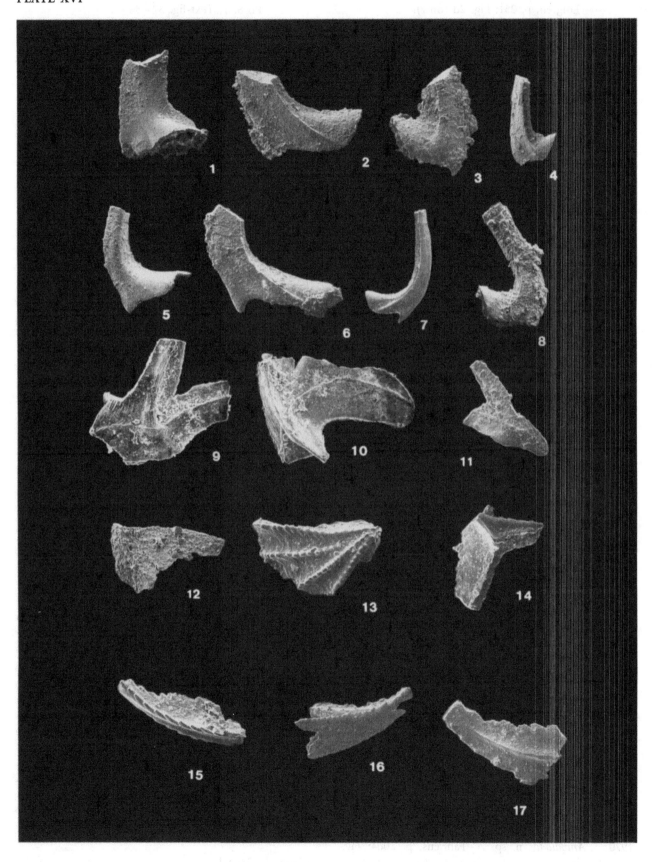

? 1990   *Polonodus* M (oistodontiform) elements –
          Löfgren, p. 251; Fig. 2d (only).
1991   *Polonodus? clivosus* (Viira) – Rasmussen,
          p. 283; Fig. 8A – C.

Description. –

Pa-1

The Pa element is stelliplanate pectiniform with four denticulated processes. The element forms an asymmetrical cross, which is characterised by an anterior process that is considerably larger than the other processes. All processes are wide with rounded, knoblike denticles situated in a median row. The upper surface of the processes is ornamented with transverse furrows and crests. The platform-margin is irregular. The central cusp is recurved. The basal cavity is deep and wide.

Pb ?

The element is pastiniplanate pectiniform. The three processes are broken close to the cusp, which only permits an inadequate description. The angle between the posterior and anterior process is about 90°. The suberect cusp is short and wide. The basal cavity is deep and wide.

M

The element is geniculate coniform with a reclined cusp. The basal margin is sinuous. The antero-basal edge is smooth. The upper margin forms an angle of 115° with the anterior margin. The basal cavity is wide.

Remarks. – Polonodus clivosus (Viira) *sensu* Dzik (1976) and Stouge (1984) differs from the type species *Ambalodus clivosus* Viira as their figured specimens lack the distinct crests and ridges orientated perpendicular to the platform ledge.

Occurrence. – Andersön-B. *B. medius - H. holodentata* Zone.

Material. – 1 Pa-1, 1 Pb?, 1 M.

Polonodus cf. *clivosus* (Viira, 1974)

Pl. 15: 9 – 10

Synonymy. –

Multielement
1970   *Polyplacognathus* n. sp. A – Fåhræus, Fig.
        3f – g.
1970   *Ambalodus* n. sp. A – Fåhræus, p. 2064; Fig.
        3j – k.

cf. 1974   *Ambalodus clivosus* n. sp. – Viira, pp. 51 – 52;
            Pl. 8: 1; Text-fig. 37 – 38.
pt. 1978   *Polonodus clivosus* (Viira) – Löfgren, p. 76;
            Pl. 16: 12a – b, 13; *non* Pl. 16: 15a – c.
1984   *Polonodus? clivosus* (Viira) – Stouge, p. 73;
            Pl. 13: 6 – 13.

Description. –

Pa

The Pa-1 ("polyplacognathiform") and Pa-2 ("ambalodontiform") elements were described in detail by Stouge (1984, p. 73).

M, Sb

The M and Sb elements seem indistinguishable from *Polonodus?* cf. *tablepointensis.* See Löfgren (1990) for description.

Remarks. – P. cf. *clivosus* differs from *P. clivosus* by lack of the characteristic numerous transverse crests and ridges on the upper surface of the Pa elements.

Occurrence. – Andersön-B. *P. graeai* Zone – *E. suecicus* Zone (basal part).

Material. – 2 Pa-1, 3 Pa-2, 3 Pa (undifferentiated), 4 M, 2 Sb.

Polonodus sp. A

Pl. 15: 11

Description. –

Pa-1

The Pa element is stelliplanate pectiniform with four denticulated processes. The angle between the outer margin of the posterior and postero-lateral processes approximates 90°. The processes are wide with rounded, knob-like denticles situated in relatively straight rows, usually close to the outer margin of the processes. The postero-lateral process may carry a median denticle row. The upper surface of the processes is ornamented with irregular knobs. The central cusp is broken in all specimens at hand. The basal cavity is deep and wide.

The species differs from the above-mentioned *Polonodus* species by a more irregular upper surface.

Occurrence. – Jøronlia, *P. graeai* Zone.

Material. – 3 Pa-1

Polonodus? cf. *tablepointensis* Stouge, 1984

Pl. 15: 1–4, 7–8, 12

Synonymy. –

Multielement
cf. 1984 *Polonodus tablepointensis* n. sp. – Stouge, p. 72; Pl. 12: 13, Pl. 13: 1–5 (*cum. syn.*).
cf. 1984 *Baltoniodus? prevariabilis medius* (Dzik) – Stouge, p. 77; Pl. 15: 1–6.
cf. 1985 *Polonodus?* sp. A Löfgren – An, Du & Gao, Pl. 17: 13, 16.
cf. 1985 *Polonodus?* sp. B Löfgren – An, Du & Gao, Pl. 17: 10.
cf. 1987 *Polonodus tablepointensis* Stouge – Hünicken & Ortega, p. 140; Pl. 7: 1–2.
? 1990 *Polonodus* elements – Löfgren, pp. 251–254; Fig. 1a–c, f–m.
? 1990 *Polonodus* M elements – Löfgren, p. 251; Fig. 2a–f.
cf. 1990 *Polonodus tablepointensis*-group P elements – Löfgren, Fig. 3a–f.
? 1991 *Eoplacognathus?* sp. A – McCracken, p. 46; Pl. 10 (only).
? 1991 *Polonodus tablepointensis* Stouge? – McCracken, p. 50; Pl. 1: 18.
    1991 *Polonodus* cf. *tablepointensis* Stouge – Rasmussen, p. 281; Fig. 7T.
cf. 1998b *Dzikodus tablepointensis* (Stouge) – Zhang, pp. 65–69; Pl. 7: 1–12; Pl. 8: 1–6.

Description. –

Pa, M, S
The Pa-1 and Pa-2 elements of *P. tablepointensis* were described by Stouge (1984, p. 72), who included two P elements in the apparatus ("polyplacognathiform" and "ambalodontiform" *sensu* Stouge) and The M and S elements correspond with the description of Löfgren (1990). Additional details and a comprehensive synonymy were given by Zhang (1998b). The material at hand indicates that the different *Polonodus* species observed in the present study have similar S and M elements.

Pb
The element is pastinate pectiniform with denticulated anterior and posterior process and a short, adenticulate lateral process. The angle between the anterior and posterior process varies between 115–130°. The cusp is suberect, relatively short, and wide. Denticles are fused close to the process, and are triangular in lateral view. The specimens at hand are small but carry about six denticles on both the anterior and posterior process. The posterior process is wide distally. The basal cavity is wide and deep with the apex placed near the anterior margin of the cusp.

Remarks. – The material presented here is similar to *P. tablepointensis* in the concentric ornamentation of the upper surface of the Pa elements, the lack of blade-like processes in Pa-1 elements, and the general outline of the elements. A confident species identification is impossible because most of the Pa elements (especially Pa-1) are badly preserved and characterised by broken or missing processes.

Occurrence. – Andersön-A, -B, Steinsodden, Røste, Jøronlia. *B. norrlandicus* - *D. stougei* Zone (uppermost part) – *E. suecicus* Zone.

Material. – 72 Pa-1, 25 Pa-2, 6 Pa (undifferentiated), 45 Pb, 15 M, 6 Sa, 25 Sb, 10 Sd.

Polonodus? sp. B

Pl. 7: 12

Description. – The element is a sinistral P with three denticulated processes. The posterior process is bifid and forms two subequal lobes that forms an angle of about 110° between them. The lateral process is slender and "blade-like". The angle between the lateral and anterior process is 155°. The anterior process has nearly the same length as the lateral process. It curves gently away from the posterior process and has an anterior situated denticle row similar to *Eoplacognathus suecicus*. The anterior process develops a short, wide, adenticulate, posterior lobe. The basal cavity is relatively narrow seen in a basal view compared to species of of the genus *Polonodus*.

Remarks. – The figured specimen resembles Pa elements of *Polonodus* Dzik. The basal cavity, however, is more slender than typical of *Polonodus*. No associated elements were found.

Occurrence. – Steinsodden. *P. graeai* Zone.

Material. – 1 P element.

Genus *Prioniodus* Pander, 1856

Type species. – *Prioniodus elegans* Pander, 1856.

Remarks. – The *Prioniodus* apparatus was reconstructed by Bergström (1968, 1971) who included six element types. Barnes *et al.* (1979, fig. 7), however, interpreted the apparatus as septimembrate comprising Pa, Pb, M, Sa, Sb, Sc and Sd elements.

Prioniodus elegans Pander, 1856

Pl. 15: 13–16

Synonymy. –

Pa

1856 *Prioniodus elegans* n. sp. – Pander, p. 29; Pl. 2: 22.

Multielement

1988 *Prioniodus elegans* Pander – Stouge & Bagnoli, pp. 133–134; Pl. 13: 1–9 (*cum. syn.*).

1988 *Prioniodus elegans* Pander – Bergström, Pl. 3: 33–38.

1995a *Prioniodus elegans* Pander – Lehnert, p. 116; Pl. 4: 4–6.

Remarks. – *Prioniodus elegans* was fully described by Stouge & Bagnoli (1988), and their description seems to agree with the few, small and badly preserved specimens at hand.

Occurrence. – Andersön-A. *P. elegans* Zone.

Material. – 2 Pa, 2 Pb, 3 M, 1 Sb.

Genus *Protopanderodus* Lindström, 1971

Type species. – *Acontiodus rectus* Lindström, 1955.

Remarks. – The apparatus of *Protopanderodus* was established by Lindström (1971) and was recently discussed in detail by McCracken (1989). *Protopanderodus* comprises three types of nongeniculate, coniform elements: scandodontiform, and symmetrical and asymmetrical acontiodontiform.

Protopanderodus calceatus Bagnoli & Stouge, 1997

Pl. 15: 20–21

Synonymy. –

Scandodontiform

cf. 1962 *Scandodus unistriatus* n. sp. – Sweet & Bergström, p. 1245; Pl. 168: 12; Text-fig. 1E.

Acontiodontiform

cf. 1962 *Scolopodus varicostatus* n. sp. – Sweet & Bergström, p. 1247; Pl. 168: 4–9; Text-fig. 1A, C, K.

Multielement

1978 *Protopanderodus* cf. *varicostatus* (Sweet & Bergström) – Löfgren, pp. 91–93; Pl. 3: 26–31.

1984 *Protopanderodus* cf. *varicostatus* (Sweet & Bergström) – Stouge, p. 51; Pl. 3: 11–17 (*cum. syn.*).

1987 *Protopanderodus* cf. *varicostatus* (Sweet & Bergström) – Olgun, Pl. 7: P – S.

1989 *Protopanderodus* cf. *varicostatus* (Sweet & Bergström) – McCracken, pp. 22–23; Pl. 3: 1–6, 7?, 8?.

1990 *Protopanderodus* cf. *varicostatus* (Sweet & Bergström) – Stouge & Bagnoli, pp. 23–24; Pl. 8: 9–12.

1991 *Protopanderodus* cf. *varicostatus* (Sweet & Bergström) – Rasmussen, pp. 283–284; Fig. 8D – E.

1994 *Protopanderodus gradatus* Serpagli – Dzik, pp. 73–74; Pl. 13: 23–26; Text-fig. 11a.

Plate XVII

1–3: *Pygodus serra* (**Hadding**).

1. Pa element, × 90. Sample 99484, Andersön-B. PMO 165.446.
2. Pb element, × 60. Sample 99484, Andersön-B. PMO 165.447.
3. Sd element, × 90. Sample 99484, Andersön-B. PMO 165.448.

4–5: *Scabbardella altipes* (**Henningsmoen**).

4. Acontiodontiform element, × 90. Sample 99510, Høyberget. PMO 165.449.
5. Acontiodontiform element, × 145. Sample 99510, Høyberget. PMO 165.450.

6–8: *Scalpellodus gracilis* (**Sergeeva**).

6. Scandodontiform element, × 95. Sample 99465, Andersön-B. PMO 165.451.
7. Shortbased drepanodontiform element, × 95. Sample 99465, Andersön-B. PMO 165.452.
8. Longbased drepanodontiform element, × 125. Sample 99465, Andersön-B. PMO 165.453.

9–11: *Scalpellodus latus* (**van Wamel**).

9. Scandodontiform element, × 95. Sample 97831, Andersön-A. PMO 165.454.
10. Shortbased drepanodontiform element, × 70. Sample 97831, Andersön-A. PMO 165.455.
11. Longbased drepanodontiform element, × 100. Sample 97831, Andersön-A. PMO 165.456.

12. *Scolopodus quadratus* **Pander.**

12. Coniform element, × 75. Sample 69687, Steinsodden. PMO 165.457.

13–14: *"Semiacontiodus" cornuformis* (**Sergeeva**).

13. Cornuform element, × 55. Sample 99465, Andersön-B. × 55. PMO 165.458.
14. Drepanodontiform element, × 85. Sample 99465, Andersön-B. PMO 165.459.

15–18: *Stolodus stola* **Lindström.**

15. P element, × 115. Sample 97805, Andersön-A. PMO 165.460.
16. Sb element, × 115. Sample 97805, Andersön-A. PMO 165.461.
17: Sc element, × 75. Sample 97805, Andersön-A. PMO 165.462.
18: Sd element, × 125. Sample 97805, Andersön-A. PMO 165.463.

PLATE XVII

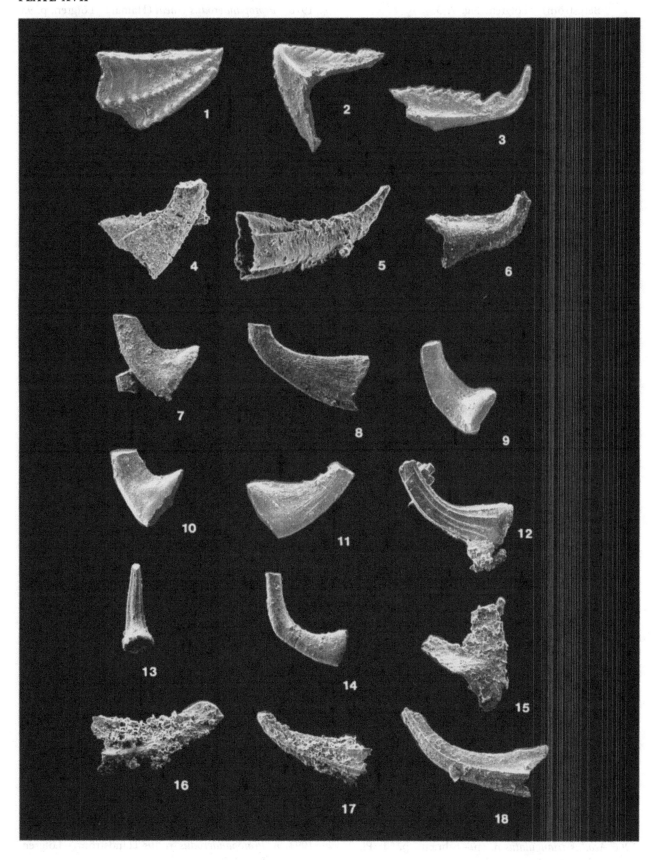

1994   *Protopanderodus* cf. *varicostatus* (Sweet & Bergström) – Löfgren, Fig. 7: 3.

1995a *Protopanderodus* cf. *varicostatus* (Sweet & Bergström) – Lehnert, p. 118; Pl. 13: 17, Pl. 17: 5 – 6, 8 – 9.

1997   *Protopanderodus calceatus* n. sp. – Bagnoli & Stouge, pp. 154 – 156; Pl. 8: 13 – 19.

1998b *Protopanderodus calceatus* Bagnoli & Stouge – Zhang, pp. 82 – 83; Pl. 15: 6 – 13.

Comments to the synonymy list. – It was suggested by Stouge & Bagnoli (1990) that *P.* cf. *varicostatus sensu* McCracken (1989) and *P.* cf. *varicostatus sensu* Löfgren (1978) and Stouge & Bagnoli (1990) probably represent two separate species.

Remarks. – All elements of this species were described by Löfgren (1978) and Bagnoli & Stouge (1997). *P. calceatus* is very similar to *Protopanderodus gradatus* Serpagli 1974, but may be distinguished by the deeper grooves on the upper part of the base and the cusp (Löfgren 1978, p. 92). It has been suggested that the two species represent two distinct lineages (Stouge 1984, McCracken 1989).

The stratigraphically old specimens previously referred to *Protopanderodus* cf. *varicostatus* (from the *O. evae* Zone) do probably belong to *P. sulcatus* (Lindström) instead of *P. calceatus* (Bagnoli & Stouge 1997).

Occurrence. – Andersön-A, -B, -C, Herram, Steinsodden, Røste, Jøronlia, Haugnes, Hestekinn, Høyberget, Engerdal. *O. evae* Zone – *P. anserinus* Zone.

Material. – 44 scandodontiform, 275 acontiodontiform elements.

Protopanderodus graeai (Hamar, 1966)

Pl. 15: 17 – 19

Synonymy. –

Symmetrical acontiodontiform
1964   *Acontiodus rectus* Lindström – Hamar, p. 258; Pl. 1: 10, 12, 13, 17; Text-fig. 4: 4a, b.

1966   *Acontiodus coniformis* n. sp. – Fåhræus, p. 15; Pl. 2: 3; Text-fig. 3b.

Scandodontiform
1966   *Acodus graeai* n. sp. – Hamar, p. 47; Pl. 3: 11 – 14; Text-fig. 3: 5.

1966   *Acodus triangulatus* n. sp. – Fåhræus, p. 11; Pl. 2: 1; Text-fig. 2c.

Multielement
1978   *Protopanderodus graeai* (Hamar) – Löfgren, p. 93; Pl. 3: 19 – 25; Text-fig. 31K – M (*cum. syn.*).

1994   *Protopanderodus graeai* (Hamar) – Dzik, pp. 72 – 73; Pl. 13: 14 – 22; Text-fig. 10b, c.

1998b *Protopanderodus graeai* (Hamar) – Zhang, pp. 84 – 85; Pl. 15: 1 – 5.

Remarks. – *Protopanderodus graeai* is characterised by long, slender, acontiodontiform elements with a proclined or suberect cusp and a deep basal cavity. The scandodontiform element has a distinct inner-lateral costa. Hamar (1966) and Löfgren (1978) described the elements of this species.

Occurrence. – Andersön-B, Steinsodden, Røste. *P. graeai* Zone – *P. anserinus-S. kielcensis* Subzone.

Material. – 41 scandodontiform, 327 symmetrical and 132 asymmetrical acontiodontiform elements.

Protopanderodus rectus (Lindström, 1955)

Pl. 16: 1 – 4

Synonymy. –

Symmetrical acontiodontiform
1955a *Acontiodus rectus* n. sp. – Lindström, p. 549; Pl. 2: 7 – 11; Text-fig. 3B.

Scandodontiform
1955a *Scandodus rectus* n. sp. – Lindström, p. 593; Pl. 4: 21 – 25; Text-fig. 3K.

Multielement
1971   *Protopanderodus rectus* (Lindström) – Lindström, p. 50.

1978   *Protopanderodus rectus* (Lindström) – Löfgren, p. 90; Pl. 3: 1 – 7, Pl. 3: 36A – B; Text-fig. 31A – C.

1981   *Protopanderodus rectus* (Lindström) – Nowlan, p. 15; Pl. 1: 6 – 7.

1987   *Protopanderodus rectus* (Lindström) – Olgun, p. 54; Pl. 7W – Z, AB, CD.

1990   *Protopanderodus rectus* (Lindström) – Stouge & Bagnoli, p. 23; Pl. 8: 1 – 5 (*cum. syn.*).

1991   *Protopanderodus rectus* (Lindström) – Rasmussen, p. 283; Fig. 8F – H.

1993b *Protopanderodus rectus* (Lindström) – Löfgren, Fig. 5A – D.

cf. 1994 *Protopande-rodus rectus* (Lindström) – Dzik, p. 72; Pl. 13: 27 – 30; Text-fig. 10a.

1994   *Protopanderodus rectus* (Lindström) – Löfgren, Fig. 7: 7 – 10.

Remarks. – The apparatus was fully described by Lindström (1955a) and Löfgren (1978). The elements figured by Dzik (1994) have a more smooth upper margin of the scandodontiform element and more pronounced basal "hooks" or heels at the anterobasal corner than typical of Baltoscandian *Protopanderodus rectus*. The species may represent an intermediate form between *P. rectus* and *P. robustus*.

Occurrence. – Andersön-A, -B, Herram, Steinsodden, Røste, Haugnes, Hestekinn, Skogstad. *O. evae* Zone – *M. ozarkodella* Zone.

Material. – 349 scandodontiform, 942 symmetrical acontiodontiform and 526 asymmetrical acontiodontiform elements.

Protopanderodus robustus (Hadding, 1913)

Pl. 16: 5 – 8

Synonymy. –

Symmetrical acontiodontiform
1913 *Drepanodus robustus* n. sp. – Hadding, p. 31; Pl. 1: 5.

Asymmetrical acontiodontiform
1964 *Distacodus* n. sp. – Hamar, p. 263; Pl. 1: 19 – 20; Text-fig. 6: 3a.

Scandodontiform
1966 *Scandodus formosus* n. sp. – Fåhræus, p. 30; Pl. 3: 11; Text-fig. 2k.

Multielement
1978   *Protopanderodus robustus* (Hadding) – Löfgren, p. 94; Pl. 3: 32 – 35; Text-fig. 31 G, J (*cum. syn.*).
1984   *Protopanderodus robustus* (Hadding) – Stouge, p. 49; Pl. 2: 3?, 4?, 5 – 8.
pt. 1989 *Protopanderodus robustus* (Hadding) – McCracken, p. 20; Pl. 1: 1 – 8 (only); Text-fig. 3E, *non* Pl. 1:9, 10.
1989   *Protopanderodus robustus* (Hadding) – Rasmussen & Stouge, Fig. 3N.
1991   *Protopanderodus robustus* (Hadding) – Rasmussen, p. 283; Fig. 8J – M.
? 1995 *Protopanderodus robustus* (Hadding) – Lehnert, p. 118; Pl. 13: 16.
1998b   *Protopanderodus cooperi* (Sweet & Bergström) – Zhang, pp. 81 – 82; Pl. 14: 13 – 17.

Comments to the synonymy list. – The scandodontiform elements illustrated by McCracken (1989, Pl. 1: 9 – 10) carry an anterior flare, which in the present author's interpretation does not occur in *Protopanderodus robustus*. The long, slender, acontiodontiform

element figured by Lehnert (1995a, Pl. 13: 16) is similar to *Protopanderodus graeai*, but the presence of a deep basal cavity must be confirmed to prove this relationship.

Remarks. – *P. robustus* resembles *P. rectus* but is distinguished from this by a smooth and more gradually rounded inner side in the scandodontiform, and a well-developed "hook" in basal margin of the symmetrical acontiodontiform. *P. rectus* was the likely ancestor of *P. robustus* (Stouge 1984). The species was described by Löfgren (1978).

Occurrence. – Andersön-A, -B, -C, Steinsodden, Røste, Jøronlia, Glöte, Høyberget, Sorken. *M. ozarkodella* Zone – *P. anserinus-A. inaequalis* Subzone.

Material. – 285 scandodontiform, 891 symmetrical acontiodontiform and 548 asymmetrical acontiodontiform elements.

Genus *Protoprioniodus* McTavish, 1973

Type species. – *Protoprioniodus simplicissimus* McTavish, 1973

Remarks. – *Protoprioniodus* McTavish was revised by Stouge & Bagnoli (1988). The apparatus comprises P, M and S elements.

Protoprioniodus cf. *costatus* (van Wamel, 1974)

Pl. 16: 11 – 12

Synonymy. –

cf. 1974 *Oelandodus costatus* n. gen. et n. sp. – van Wamel, pp. 72 – 74; Pl. 7: 5 – 7.
? 1988 *Oelandodus? costatus* van Wamel – Bagnoli & Stouge, pp. 210 – 211; Pl. 39: 9, 10, 12.
? 1994 *Protoprioniodus* sp. B. – Löfgren, Fig. 8: 36.

Description. –

P
The element is carminate pectiniform with adenticulate anterior and posterior processes. The cusp is strongly reclined and the angle between the posterior margin of the cusp and the upper margin of the base is about 35°. The cusp develops a median carina with distinct striae. The upper margin of the anterior process and the anterior margin of the cusp form an almost straight line. The lower part of the element is expanded from the anterior to the posterior tip, and

forms a shelf at the bottom of the cusp. The basal cavity is narrow and shallow and extends the entire length of the element.

Sb

The Sb element is identical with the *Protoprioniodus* sp. A elements (see description of this).

Remarks. – Van Wamel (1974) described and illustrated the P ("oistodontiform"), M ("elongatiform") and Sa ("triangulariform") elements of *"Oelandodus" costatus*.

The present P element is similar to the corresponding element of *Protoprioniodus costatus*. However, a confident assignment of the material at hand to *P. costatus* is difficult because van Wamel (1974) did not illustrate the Sb element, and because no M or Sa elements were recognised in the present material.

"Oelandodus?" costatus sensu Bagnoli & Stouge (1988) includes a P element with a curved anterior margin instead of a straight, and thus is only questionably referred to this species.

The element assigned to *Protoprioniodus* sp. B by Löfgren (1994) is possibly conspecific with *P.* cf. *costatus*, but the picture (Löfgren 1994, Fig. 8:36) does not show whether or not the anterior part is striated.

Occurrence. – Andersön-A. *P. proteus* Zone.

Material. – 1 P and 2 Sb elements.

Protoprioniodus sp. A

Pl. 16: 9 – 10

Synonymy. –

P

pt. aff. 1988 *Protoprioniodus simplicissimus* McTavish – Stouge & Bagnoli, p. 138; Pl. 14: 7 (only).

S

pt. aff. 1988 *Protoprioniodus cowheadensis* n. sp. – Stouge & Bagnoli, p. 137; Pl. 14: 5 (only).

Description. –

P

The element is carminate pectiniform with adenticulate anterior and posterior processes. The cusp is reclined and the angle between the posterior margin of the cusp and the upper margin of the base is about 60°. The cusp develops a median carina with distinct striae. The lower part of the element is expanded from the anterior to the posterior tip, thus forming a shelf at the bottom of the cusp. The basal cavity is narrow

and shallow and extends along the entire length of the element.

Sb

The element is tertiopedate ramiform with adenticulate anterior and posterior processes. The cusp is suberect with an outer-lateral costa that continues downward into an adenticulate, lateral process. The lower part of the element is expanded forming a small, longitudinal shelf that extends from the posterior tip to the tip of the anterior process. The anterior margin of the element is striated above the shelf with striae situated parallel to the lateral process.

Remarks. – The elements are very similar to the abovementioned elements illustrated by Stouge & Bagnoli (1988) but are distinguished from these by a wider expanded area in the lower part of the element.

Occurrence. – Herram. *M. flabellum - D. forceps* Zone.

Material. – 3 P and 2 Sb elements.

Genus *Pygodus* Lamont & Lindström, 1957

Type species. – *Pygodus anserinus* Lamont & Lindström, 1957

Remarks. – The apparatus was considered bi- or probably tetramembrate by Bergström (1971, 1983) but recently it was shown that the apparatus likely is seximembrate and comprises Pa (stelliscaphate pectiniform), Pb (pastinate pectiniform), Sa (alate ramiform), Sb (asymmetrical tertiopedate ramiform), Sc (asymmetrical bipennate ramiform) and Sd (quadriramate ramiform) elements (McCracken 1991).

Hitherto, five multielement species of the *Pygodus*-lineage have been formally described (Löfgren 1978; Bergström 1983; Zhang 1998a), of which *P. serra* (Hadding 1913) and *P. anserinus* Lamont & Lindström 1957 are well known world-wide. The ancestor species is *P. anitae* Bergström 1983 (*Pygodus* sp. C of Löfgren 1978). It is possible that *Pygodus?* n. sp. Bergström 1983 (= *Pygodus?* sp. B Löfgren 1978) represents the oldest species of the *Pygodus*-lineage, although the Pa elements they were associated with by Zhang (1998a, Pl. 1:12, 13) in the species *Pygodus lunnensis* Zhang are more similar to *Polonodus* in the present author's opinion.

Pygodus anserinus Lamont & Lindström, 1957

Pl. 16: 13 – 17

Pa

1957 *Pygodus anserinus* n. sp. – Lamont & Lindström, p. 67; Pl. 5: 12, 13; Text-fig. 1a – d.

1960 *Pygodus anserinus* Lamont & Lindström – Lindström, p. 91, 95; Fig. 7: 3.

1962 *Pygodus anserinus* Lamont & Lindström – Sweet & Bergström, p. 1241; Pl. 171: 11, 12; Text-fig. 4.

1964 *Pygodus anserinus* Lamont & Lindström – Hamar, p. 279; Pl. 4: 1 – 4, 11.

1964 *Pygodus anserinus* Lamont & Lindström – Lindström, Fig. 56j.

1966 *Pygodus anserinus* Lamont & Lindström – Hamar, Pl. 7: 1.

1969 *Pygodus* sp. – Ethington & Schumacher, p. 475; Pl. 69: 16, 19.

1974 *Pygodus anserinus* Lamont & Lindström – Viira, p. 115; Pl. 11: 26, 27.

Pb

1962 *Haddingodus serra* (Hadding) – Sweet & Bergström, p. 1229; Pl. 170: 1, 4.

1964 *Haddingodus serra* (Hadding) – Hamar, p. 266; Pl. 4: 13, 16; Text-fig 5: 6a – b.

Sa

1962 *Roundya pyramidalis* n. sp. – Sweet & Bergström, p. 1243; Pl. 170: 7 – 9.

1964 *Roundya pyramidalis* Sweet & Bergström – Hamar, p. 280; Pl. 5: 15, 16, 20, 21; Text-fig. 4: 12.

Sd

1962 *Tetraprioniodus lindstroemi* n. sp. – Sweet & Bergström, p. 1248; Pl. 170: 5, 6.

1964 *Tetraprioniodus lindstroemi* Sweet & Bergström – Hamar, p. 291; Pl. 6: 4, 5; Text-fig. 4: 14.

Multielement

1971 *Pygodus anserinus* Lamont & Lindström – Bergström, p. 149; Pl. 2: 20, 21.

1974 *Pygodus anserinus* Lamont & Lindström – Bergström, Riva & Kay, p. 1644, 1646; Pl. 1: 16, 17.

1976 *Pygodus anserinus* Lamont & Lindström – Dzik, Fig. 29f.

1978 *Pygodus anserinus* Lamont & Lindström – Bergström, Pl. 79: 1 – 2.

1979 *Pygodus anserinus* Lamont & Lindström – Harris, Bergström, Ethington and Ross, Pl. 3: 16, 17, Pl. 4: 17.

pt. 1981 *Pygodus* cf. *P. serrus* (Hadding) – Nowlan, Pl. 2: 18 – 20 (only).

1985 *Pygodus anserinus* Lamont & Lindström – Bergström & Orchard, p. 58; Pl. 2.3: 3.

1985 *Pygodus anserinus* Lamont & Lindström – An, Du & Gao, Pl. 17: 7 – 9.

1989 *Pygodus anserinus* Lamont & Lindström – Rasmussen & Stouge, Fig. 3a – c.

1990 *Pygodus anserinus* Lamont & Lindström – Männik & Viira, Pl. 16: 32.

1990 *Pygodus anserinus* Lamont & Lindström – Bergström, Pl. 1: 19 – 22.

1990 *Pygodus anserinus* Lamont & Lindström – Pohler & Orchard, Pl. 2: 20.

1994 *Pygodus anserinus* Lamont & Lindström – Dzik, pp. 105 – 106; Pl. 17: 7 – 8; Textfigs. 26 – 27.

1995a *Pygodus anserinus* Lamont & Lindström – Lehnert, pp. 120 – 121; Pl. 17: 16, 18.

1997 *Pygodus anserinus* Lamont & Lindström – Armstrong, pp. 777 – 778; Pl. 4: 1 – 7, Text-fig. 4.

1998 *Pygodus anserinus* Lamont & Lindström – Bednarczyk, Pl. 2: 4, 16.

1998a *Pygodus anserinus* Lamont & Lindström – Zhang, Pl. 3: 1 – 8; Text-fig. 2E.

1998b *Pygodus anserinus* Lamont & Lindström – Zhang, pp. 87 – 88; Pl. 16: 1 – 5.

Description. – The Pa element was described by Lamont & Lindström (1957) and the Pb, Sa and Sd elements by Sweet & Bergström (1962). No Sc elements were observed, although such elements have been described from the ancestor *Pygodus serra* (McCracken 1991).

Sb

The element is tertiopedate ramiform with denticulated posterior and anterior processes. The outer-lateral process is adenticulate with a sharp upper margin. The angle between the anterior and posterior process is about 15°. The cusp and denticles are proclined.

Remarks. – It seems impossible to distinguish between *Pygodus serra* and *P. anserinus* with regard to the S elements.

Transitional Pa elements between *P. serra* and *P. anserinus* develop scattered nodes arranged in a straight line instead of a true fourth denticle row (Nowlan 1981). Such elements have been included in *Pygodus anserinus* within the present study.

The angle between the anterior and posterior process in Pb elements usually vary between 70° and 90°. This is slightly more than typical of *Pygodus serra*, which has an angle of about 70°.

Occurrence. – Andersön-B, -C, Glöte, Høyberget, Engerdal. *P. anserinus-S. kielcensis* Subzone – *P. anserinus-A. inaequalis* Subzone.

Material. – 192 Pa, 112 Pb, 10 Sa, 21 Sb, 1 Sc, 10 Sd.

Pygodus serra (Hadding, 1913)

Pl. 17: 1 – 3

Pa

1960   *Pygodus* n. sp. 2 – Lindström, p. 91; Fig. 7: 1.

1961   *Pygodus anserinus* Lamont & Lindström – Wolska, p. 357; Figs. 4, 5.

1966   *Pygodus trimontis* n. sp. – Hamar, p. 70; Pl. 7: 12, 16, 17.

1967   *Pygodus* aff. *anserinus* Lamont & Lindström – Viira, Fig. 4: 6.

Pb

1913   *Arabellites serra* n. sp. – Hadding, p. 33; Pl. 1: 12, 13.

1967   *Haddingodus serra* (Hadding) – Viira, Fig. 4: 7.

1974   *Haddingodus serra* (Hadding) – Viira, p. 86; Pl. 11: 25; Text-fig. 105.

Multielement

1971   *Pygodus serrus* (Hadding) – Bergström, p. 149; Pl. 2: 22, 23.

1974   *Pygodus serrus* (Hadding) – Bergström, Riva & Kay, Pl. 1: 18.

pt. 1976   *Pygodus serrus* (Hadding) – Dzik, Fig. 29a, b (only).

1978   *Pygodus serra* (Hadding) – Löfgren, p. 98; Fig. 32D – F.

1979   *Pygodus serra* (Hadding) – Harris *et al.*, Pl. 2: 18.

pt. 1981   *Pygodus* cf. *P. serrus* (Hadding) – Nowlan, Pl. 2: 14, 16, 17 (only).

1985   *Pygodus serra* (Hadding) – Bergström & Orchard, p. 56; Pl. 2.2: 5.

1985   *Pygodus serrus* (Hadding) – An, Du & Gao, Pl. 17: 2 – 6.

1990   *Pygodus serra* (Hadding) – Bergström, Pl. 1: 23, 24.

1990   *Pygodus serra* (Hadding) – Männik & Viira, Pl. 16: 31.

1990   *Pygodus serra* (Hadding) – Pohler & Orchard, Pl. 1: 18.

1991   *Pygodus serra* (Hadding) – McCracken, p. 51; Pl. 2: 4, 6, 7, 9, 11, 12, 14 – 18, 20 – 23, 28 – 30.

1994   *Pygodus serra* (Hadding) – Dzik, pp. 103 – 105; Pl. 17: 9 – 12; Text-fig. 26.

1995a   *Pygodus serra* (Hadding) – Lehnert, p. 121; Pl. 17: 14.

1998a   *Pygodus serra* (Hadding) – Zhang, p. 96; Pl. 2: 1 – 14; Fig. 2C.

1998a   *Pygodus protoanserinus* n. sp. – Zhang, pp. 96 – 97; Pl. 3: 9 – 18; Fig. 2D.

1998b   *Pygodus protoanserinus* Zhang – Zhang, pp. 86 – 87; Pl. 16: 6 – 8.

Remarks. – The *Pygodus serra* Pa element is separated from that of *P. anserinus* by having three denticle rows on the upper surface instead of four, whereas the Pb element has a more narrow angle between the anterior and posterior processes (70°) than typical of *P.*

anserinus (70 – 90°). The Pa and Pb elements were described by Bergström (1971, p. 150) and the ramiform elements by McCracken (1991). Zhang (1998a) distinguished *Pygodus protoanserinus* from *P. serra* on differences in the curvature and the relative distances between the denticle rows in the Pa ("pygodontiform") elements. The two species were shown to have overlapping ranges and similar Pb and S elements at the Lunne section in Jämtland (Zhang 1998a), and this is also the case in the allochthonous areas further west. Therefore, "*P. protoanserinus*" is here considered as a morphological variation of *Pygodus serra*. Similar differences in the curvature and the relative distances between the denticle rows in the Pa elements are also known from *Pygodus anserinus*.

Occurrence. – Andersön-B. *P. serra* - *E. reclinatus* Subzone.

Material. – 8 Pa, 13 Pb, 1 Sb, 2 Sd.

Genus *Scabbardella* Orchard, 1980

Type species. – *Drepanodus altipes* Henningsmoen, 1948

Remarks. – The *Scabbardella* apparatus was reconstructed and described by Orchard (1980) who recognised three principal element-types representing six morphotypes (two drepanodontiform, two acodontiform and two distacodontiform elements).

Plate XVIII

1: *Strachanognathus parvus* Rhodes.
1. ?P element, × 105. Sample 99465, Andersön-B. PMO 165.464.

2–10: *Tetraprioniodus robustus* Lindström.
2: Pa element, × 75. Sample 97799, Andersön-A. PMO 165.465.
3: Pa element, × 100. Sample 97799, Andersön-A. PMO 165.466.
4: Pa element, × 90. Sample 97799, Andersön-A. PMO 165.467.
5: Pb element, × 145. Sample 97799, Andersön-A. PMO 165.468.
6: M element, × 60. Sample 97799, Andersön-A. PMO 165.469.
7: Sa element, × 130. Sample 99452, Andersön-B. PMO 165.470.
8: Sb element, × 100. Sample 97799, Andersön-A. PMO 165.471.
9: Sc element, × 90. Sample 97799, Andersön-A. PMO 165.472.
10: Sd element, × 65. Sample 97799, Andersön-A. PMO 165.473.

11: Gen. et sp. indet. A.
11. ?Pa element, × 110. Sample 99452, Andersön-B. PMO 165.474.

12: Gen. et sp. indet. B.
12. Scandodontiform element, × 50. Sample 99479, Andersön-B. PMO 165.475.

13: Gen. et sp. indet. D.
13. Pa element, × 80. Sample 99464, Andersön-B. PMO 165.476.

14–15: Gen. et sp. indet. C.
14. ?P element, × 230. Sample 69616, Steinsodden. PMO 165.477.
15. Sd element, × 115. Sample 69616, Steinsodden. PMO 165.478.

PLATE XVIII

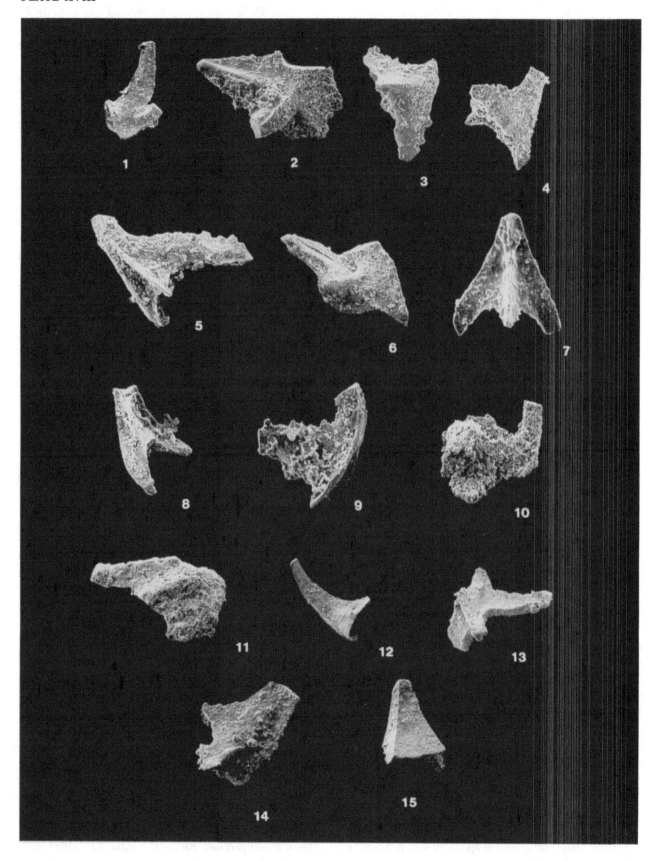

Scabbardella altipes Henningsmoen, 1948

Pl. 17: 4–5

Synonymy. –

Acostate drepanodontiform
1948　*Drepanodus altipes* n. sp. – Henningsmoen in
　　　Waern *et al.*, p. 420; Pl. 25: 14.

Acodontiform
1985　*Dapsilodus similaris* (Rhodes) – An, Du & Gao, Pl.
　　　11: 9, 10, 13, ?14.

Multielement
1980　*Scabbardella altipes* n. ssp. A – Orchard, pp. 25–
　　　26; Pl. 5: 4, 5, 7, 8, 12, 28, 35 (*cum. syn.*).
1983　*Scabbardella altipes* (Henningsmoen) – Nowlan,
　　　pp. 667–668; Pl. 1: 6, 7, 11–14.
1985　*Scabbardella altipes* (Henningsmoen) – Berg-
　　　ström & Orchard, p. 64; Pl. 2.5: 13.
1989　*Scabbardella altipes* (Henningsmoen) – Ras-
　　　mussen & Stouge, Figs. 3M, P, Q.
1990　*Scabbardella altipes* (Henningsmoen) – Berg-
　　　ström, Pl. 4: 14.
1994　*Scabbardella altipes* (Henningsmoen) – Dzik,
　　　pp. 64–66; Pl. 11: 36–39; Text-fig. 6e.
1998b *Scabbardella altipes* (Henningsmoen) – Zhang,
　　　p. 89; Pl. 17: 1–4.

Remarks. – The elements of *S. altipes* were described in
detail by Orchard (1980, pp. 25–26).

Occurrence. – Høyberget, Engerdal. *P. anserinus-A. inaequalis* Subzone.

Material. – 6 drepanodontiform, 26 acodontiform, 1
distacodontiform.

Genus Scalpellodus Dzik, 1976

Type species. – *Protopanderodus latus* van Wamel, 1974

Remarks. – The apparatus is trimembrate and consists of
longbased drepanodontiform, shortbased drepanodontiform and scandodontiform elements (Löfgren 1978).

Scalpellodus gracilis (Sergeeva, 1974)

Pl. 17: 6–8

Synonymy. –

pt. 1974　*Scandodus gracilis* n. sp. – Sergeeva, p. 80; Pl. 1:
　　　　6–9 (only).

1978　*Scalpellodus gracilis* (Sergeeva) – Löfgren, p. 100;
　　　Pl. 5: 3–6, Pl. 5: 11–13, 15, Pl. 6: 5–6, 8–20,
　　　22–23 (with additional synonymy).
1990　*Scalpellodus gracilis* (Sergeeva) – Stouge &
　　　Bagnoli, pp. 24–25; Pl. 8: 20–29 (*cum. syn.*).
1991　*Scalpellodus gracilis* (Sergeeva) – Rasmussen,
　　　p. 284; Fig. 8P, Q, V.

Remarks. – The present material adds nothing new to
the description by Löfgren (1978).

Occurrence. – Andersön-A, -B, Steinsodden, Røste,
Haugnes, Skogstad. *B. norrlandicus* - *D. stougei*
Zone – *E. suecicus* Zone.

Material. – 74 scandodontiform, 308 longbased drepanodontiform and 148 shortbased drepanodontiform
elements.

Scalpellodus latus (van Wamel, 1974)

Pl. 17: 9–11

Synonymy. –

1974　　　*Protopanderodus latus* n. sp. – van Wamel,
　　　　　p. 91; Pl. 4: 1–3.
1978　　　*Scalpellodus latus* (van Wamel) – Löfgren,
　　　　　p. 99; Pl. 5: 10, 14, Pl. 6: 1–4, 7, 21.
non 1981　*Scalpellodus latus* (van Wamel) – Cooper,
　　　　　p. 179; Pl. 27: 7–10, 13–15.
1985　　　*Scalpellodus latus* (van Wamel) – Löfgren,
　　　　　Fig. 4AL–AP.
1990　　　*Scalpellodus latus* (van Wamel) – Stouge &
　　　　　Bagnoli, p. 25; Pl. 8: 13–19.
1991　　　*Scalpellodus latus* (van Wamel) – Rasmus-
　　　　　sen, p. 284; Fig. 8I, N, O.
non 1995a *Scalpellodus latus* (van Wamel) – Lehnert,
　　　　　pp. 122–123; Pl. 12: 8, 22.
1995b　　*Scalpellodus latus* (van Wamel) – Löfgren,
　　　　　Fig. 9ae–ag.

Remarks. – *Scalpellodus latus* (van Wamel) was
described in detail by Löfgren (1978). The elements
illustrated by Cooper (1981) bear much more distinct
keels at the drepanodontiform elements than typical of
S. latus and do not belong to this species. Stouge &
Bagnoli (1990) suggested that *S. latus sensu* Cooper
(1981) may belong to "*Semiacontiodus*" *cornuformis
sensu lato*.

Occurrence. – Andersön-A, -B, Steinsodden. *M.
flabellum*-uppermost part) – *B. medius – H. holodentata*
Zone.

Material. – 73 scandodontiform, 329 longbased drepanodontiform and 154 shortbased drepanodontiform elements.

Genus *Scolopodus* Pander, 1856

Type species. – *Scolopodus sublaevis* Pander, 1856

Scolopodus quadratus Pander, 1856

Pl. 17: 12

Synonymy. –

1856　*Scolopodus quadratus* n. sp. – Pander, p. 26; Pl. 2: 6a – d, Pl. A: 5d.
1856　*Scolopodus costatus* n. sp. – Pander, p. 26; Pl. 2: 7a – d, Pl. A: 5e.
1856　*Scolopodus striatus* n. sp. – Pander, p. 26; Pl. 2: 8a – d, Pl. A: 5f.
1955a　*Scolopodus rex* n. sp. – Lindström, p. 595; Pl. 3: 32.
1955a　*Scolopodus rex* n. sp. var. *paltodiformis* nov. – Lindström, p. 596; Pl. 3: 33, 34.
1978　*Scolopodus rex* Lindström – Löfgren, p. 109; Pl. 1: 38 – 39 (*cum. syn.*).
1982　*Scolopodus quadratus* Pander – Fåhræus, p. 21; Pl. 2: 1 – 14, Pl. 3: 1 – 8, 15.
1987　*Scolopodus quadratus* Pander – Olgun, Pl. 7EF – RS.
1988　*Scolopodus* spp. – Bergström, Pl. 3: 43 – 45.
1990　*Scolopodus rex* Lindström, Stouge & Bagnoli, p. 25; Pl. 9: 1 – 6.
1994　*Scolopodus rex* Lindström – Löfgren, Fig. 7: 1.
1995b　*Scolopodus? rex* Lindström – Löfgren, Fig. 9as – at.
1998b　*Scolopodus rex* (Lindström) – Zhang, pp. 90 – 91; Pl. 17: 5 – 8.

Remarks. – *Scolopodus* Pander was revised and described by Fåhræus (1982), who interpreted *Scolopodus rex* Lindström as a junior synonym of *Scolopodus quadratus* Pander.

Occurrence. – Andersön-A, -B, Steinsodden. *P. proteus* Zone – *M. flabellum – D. forceps* Zone.

Material. – 20 multicostate coniform specimens.

"*Scolopodus*" *peselephantis* Lindström, 1955

Not figured

Synonymy. –

1955a　*Scolopodus? peselephantis* n. sp. – Lindström, p. 95; Pl. 2: 19, 20; Text-fig. 3Q.
1978　*Scolopodus? peselephantis* Lindström – Löfgren, p. 108; Pl. 4: 43 – 47 (*cum. syn.*).
1982　"*Scolopodus*" *peselephantis* Lindström – Ethington & Clark, p. 102; Pl. 11: 26.
1985　*Scolopodus? peselephantis* Lindström – An, Du & Gao, Pl. 10: 6, 7.
1988　"*Scolopodus*" *peselephantis* Lindström – Stouge & Bagnoli, p. 139; Pl. 15: 18.
1990　"*Scolopodus*" *peselephantis* Lindström – Stouge & Bagnoli, p. 25; Pl. 9: 12, 13.
1994　*Scolopodus peselephantis* Lindström – Dzik, p. 60; Pl. 11: 1 – 5.
1994　"*Scolopodus*" *peselephantis* Lindström – Löfgren, Fig. 7: 2.
1995b　"*Scolopodus*" *peselephantis* Lindström – Löfgren, Fig. 7am.
1997b　"*Scolopodus*" *peselephantis* Lindström – Löfgren, Fig. 4C.
pt. 1997　*Strachanognathus parvus* Rhodes – Armstrong, p. 789 – 791; Pl. 5: 4 – 5 (only); Text-fig. 8qa, qg (only).
1998b　*Scolopodus peselephantis* Lindström – Zhang, pp. 89 – 90; Pl. 17: 9 – 10.

Remarks. – "*Scolopodus*" *peselephantis* forms a symmetry transition series (Stouge & Bagnoli 1990). The present specimens of "*Scolopodus*" *peselephantis* agree with the descriptions by Lindström (1955a) and Stouge & Bagnoli (1990).

Occurrence. – Andersön-A, -B, Herram, Steinsodden, Røste, Grøslii, Haugnes, Skogstad. *P. deltifer* Zone – *E. suecicus* Zone.

Material. – 67 specimens.

Genus *Semiacontiodus* Miller, 1969

Type species. – *Acontiodus (Semiacontiodus) nogamii* Miller, 1969

Remarks. – *Semiacontiodus* was discussed and revised by Miller (1980), and from that it seems likely that "*Semiacontiodus*" *cornuformis* is not a true *Semiacontiodus*. The apparatus is basically bimembrate and comprises cornuform and drepanodontiform elements (Dzik 1976; Löfgren 1978).

"Semiacontiodus" cornuformis (Sergeeva, 1963)

Pl. 17: 13–14

Synonymy. –

1963   *Scolopodus cornuformis* n. sp. – Sergeeva,
            p. 93; Pl. 7: 1–3; Text-figs. 1.
1978   *Scolopodus cornuformis* Sergeeva – Löf-
            gren, p. 105; Pl. 7: 1–6, 9–12, Pl. 8:
            1–2, 4–6 (*cum. syn.*).
1990   *"Semiacontiodus" cornuformis* (Ser-
            geeva) – Stouge & Bagnoli, p. 26; Pl. 9:
            14–18, 20–25.
1991   *Semiacontiodus cornuformis* (Sergeeva) –
            Rasmussen, pp. 284–285.
1994   *Semiacontiodus cornuformis* (Sergeeva) –
            Dzik, pp. 66–67; Pl. 13: 7–10; Text-
            fig. 7a.
? pt. 1995a *Semiacontiodus cornuformis* (Sergeeva) –
            Lehnert, pp. 125–126; Pl. 9:21–22
            (only), Pl. 12: 19, 21, 24 (only).
non pt. 1995a *Semiacontiodus cornuformis* (Sergeeva) –
            Lehnert, pp. 125–126; Pl. 7: 22, Pl. 9: 14
            (only), Pl. 12: 18, 23 (only).
1997   *Semiacontiodus cornuformis* (Sergeeva) –
            Bagnoli & Stouge, p. 159.

Remarks. – The present specimens fit with the descriptions of Löfgren (1978) and Stouge & Bagnoli (1990). The specimens figured by Lehnert (1995a, Pl. 7:22, Pl. 9: 14, and Pl. 12: 18, 23) carry a median lateral groove, which does not occur in *"Semiacontiodus" cornuformis*.

Stouge & Bagnoli (1990, p. 26) distinguished three different cornuform morphotypes of *"Semiacontiodus" cornuformis*, which have been designated cornuform A, B and C in the present paper. The stratigraphically oldest morphotype (A) is characterised by a cornuform element with two vaguely developed postero-lateral grooves. The succeeding morphotype (B) is a cornuform element with two distinct, postero-lateral grooves, and a median groove on the posterior carina. The youngest morphotype (C) (*"S." cornuformis sensu stricto*) also has two well-defined postero-lateral grooves on the cornuform element, but lacks the groove on the median posterior carina.

The morphotype C specimens obtained from the Andersön and Røste sections are characterised by a sharp, posterior keel, which have not been recognised on the specimens from the Stein section. The presence of the keel is not related to stratigraphy.

It is indeed possible that future studies will show that the A, B and C morphotypes should be considered as separate species as they are stratigraphically separated both on

Öland (Stouge & Bagnoli 1990) and near the Baltoscandian platform edge (this study).

Occurrence. – Andersön-A, -B, Steinsodden, Røste, Haugnes, Glöte. *P. rectus* – *M. parva* Zone (upper part) – *P. anserinus* Zone. Its range within the Lower Allochthon is from the *P. rectus* – *M. parva* Zone to the *P. graeai* Zone.

Material. – 53 cornuform (12 A, 4 B, 37 C) and 196 drepanodontiform elements.

Genus *Stolodus* Lindström, 1971

Type species. – *Distacodus stola* Lindström, 1955

Discussion. – The *Stolodus* apparatus was considered unimembrate by Lindström (1971) in his multielement reconstruction, and this was followed by Bergström (in Robison 1981). Van Wamel (1974), however, interpreted the genus as trimembrate and included the form – species *Distacodus stola* var. *latus* Lindström in the apparatus. *Distacodus stola* s. f. and similar elements with one anterior and one posterior keel and between one and three lateral costae are here interpreted as S elements, while *D. stola* var. *latus* s. f. is tentatively interpreted as the P element. The restricted material at hand has revealed that the apparatus is at least quadrimembrate consisting of one P element and three S element types making a symmetry transition-series due to the location of costae.

Plate XIX

1–7: *Triangulodus amabilis* n. sp.
1. P element, × 80. Sample 69687, Steinsodden. PMO 165.479.
2. P element, × 70. Sample 69687, Steinsodden. PMO 165.480.
3. M element, × 60. Sample 69687, Steinsodden. PMO 165.481.
4. Sa element, × 105. Sample 69687, Steinsodden. PMO 165.482.
5. Sb element, × 60. Sample 69687, Steinsodden. PMO 165.483.
6. Sc element, × 60. Sample 69687, Steinsodden. PMO 165.484.
7. Sd element, × 120. Sample 69687, Steinsodden. PMO 165.485. Holotype.

8–11: *Triangulodus brevibasis* (Sergeeva).
8. M element, × 125. Sample 99458, Andersön-B. PMO 165.486.
9. Sc element, × 65. Sample 99458, Andersön-B. PMO 165.487.
10. Sb element, × 145. Sample 99458, Andersön-B. PMO 165.488.
11. Sd element, × 70. Sample 99458, Andersön-B. PMO 165.489.

12–17: *Trapezognathus quadrangulum* Lindström
12. Pa element, × 100. Sample 69685, Herram. PMO 165.490.
13. Pb element, × 100. Sample 69685, Herram. PMO 165.491.
14. M element, × 110. Sample 69685, Herram. PMO 165.492.
15. Sa element, × 125. Sample 69685, Herram. PMO 165.493.
16. Sc element, × 125. Sample 69685, Herram. PMO 165.494.
17. Sd element, × × 170. Sample 69685, Herram. PMO 165.495.

PLATE XIX

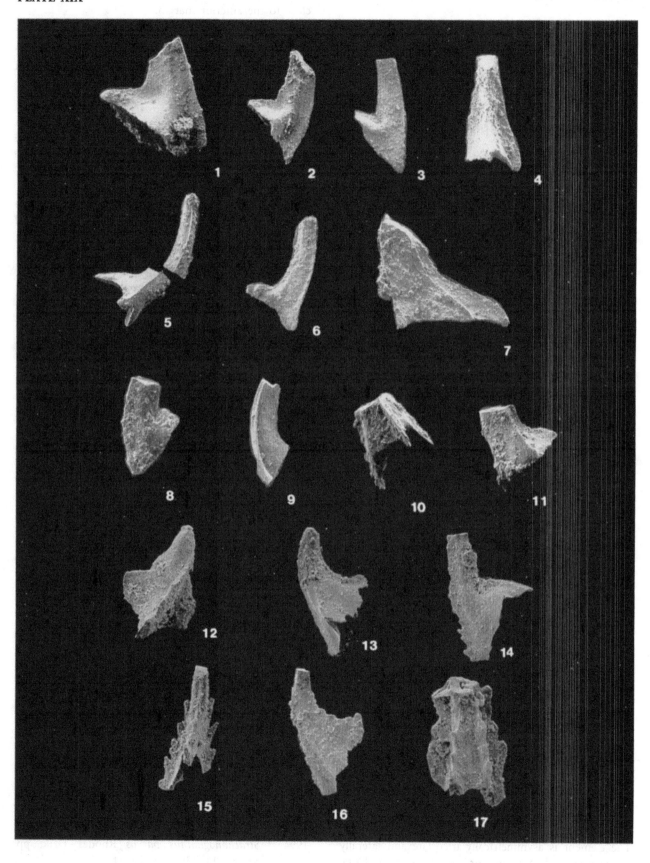

Stolodus stola (Lindström, 1955)

Pl. 17: 15 – 18

Synonymy. –

P
1955a *Distacodus stola* n. sp. var. *latus* nov. – Lindström, p. 557; Pl. 3: 50 – 51.
1974   *Distacodus latus* Lindström – Viira, Pl. 3: 16.

S
1955a   *Distacodus stola* n. sp. – Lindström, p. 556; Pl. 3: 43 – 49.
1960   *Distacodus stola* Lindström – Lindström, Fig. 2: 2.
1961   *Distacodus stola* Lindström – Wolska, p. 348; Pl. 2: 4.
1964   *Coelocerodontus stola* (Lindström) – Lindström, Fig. 29.
non 1971 *Distacodus stola* Lindström – Ethington & Clark, Pl. 2: 16.
1974   *Distacodus stola* Lindström – Viira, Pl. 3: 17.

Multielement
1971   *Stolodus stola* (Lindström) – Lindström, p. 51.
pt. 1974 *Stolodus stola* (Lindström) – van Wamel, p. 95; Pl. 8: 20 – 22, 24, *non* Pl. 8: 23.
1978   *Stolodus stola stola* (Lindström) – Löfgren, p. 111; Pl. 9: 18 – 19.
1978   *Stolodus stola latus* (Lindström) – Löfgren, p.112; Pl. 9: 20 – 23.
1982   *Stolodus stola* (Lindström) – Ethington & Clark, p. 108; Pl. 12: 21.
1987   *Stolodus stola latus* (Lindström) – Olgun, Pl. 6E, F.
pt. 1987 *Stolodus stola stola* (Lindström) – Olgun, p. 55; Pl. 7TU – XY, *non* ZZ (= Gen. et sp. indet. C).
1988   *Stolodus stola* (Lindström) – Bergström, Pl. 3: 46.
1993b   *Stolodus stola* (Lindström) – Löfgren, Fig. 6F – H.

Description. –

P
The P element was described by Lindström (1955a, p. 557) as the form species *Distacodus stola* var. *latus*.

S elements in general
The elements are adenticulate, nongeniculate coniform, forming a symmetry transition series with respect to the number and placement of costae. The costae are sharp. The cusp is proclined. The basal cavity is deep with the apex near the anterior margin at the point of maximum element curvature. Walls are thin. The base is sometimes yellowish in this material.

Sb
The element is asymmetrically tricostate and laterally compressed with one anterior, one posterior and one outer-lateral costa. The outer-lateral costa is situated close to the anterior margin.

Sc
The element is quadricostate and weakly laterally compressed. The costae are situated in a asymmetrical pattern with 2 posterior and 2 anterior costae.

Sd
The element is quadricostate and laterally compressed, characterised by one anterior, one posterior and two lateral costae. The element is slightly asymmetrical.

Remarks. – The denticulated morphotype assigned to *Stolodus stola* by van Wamel (1974, Pl. 8: 23) and Bergström & Klapper (in Robison 1981) has not been recognised within the present material. The element was excluded from *S. stola* by Stouge & Bagnoli (1988) and that interpretation is followed here.

Occurrence. – Andersön-A, -B, Herram. *P. elegans* Zone – *M. flabellum – D. forceps* Zone.

Material. – 3 P, 4 Sb, 12 Sc, 19 Sd.

Genus *Strachanognathus* Rhodes, 1955

Type species. – *Strachanognathus parvus* Rhodes, 1955

Remarks. – The apparatus forms a curvature symmetry transition series and has been interpreted as seximembrate (Dzik 1991, Fig. 7D) and quinquemembrate (Armstrong 1997, Text-fig. 8).

Strachanognathus parvus Rhodes, 1955

Pl. 18: 1

Synonymy. –

1955   *Strachanognathus parvus* gen. et sp. nov. – Rhodes, p. 132; Pl. 7: 16, Pl. 8: 1 – 4.
1978   *Strachanognathus parvus* Rhodes – Löfgren, p. 112; Pl. 1: 29 (*cum. syn.*).
1984   *Strachanognathus parvus* Rhodes – Stouge, p. 57; Pl. 5: 9.
1985   *Strachanognathus parvus* Rhodes – Bergström & Orchard, p. 58; Pl. 2.3: 1.
1991   *Strachanognathus parvus* Rhodes – McCracken, p. 52; Pl. 2: 36 (*cum. syn.*).
1991   *Strachanognathus parvus* Rhodes – Dzik, Fig. 7D.
1994   *Strachanognathus parvus* Rhodes – Dzik, pp. 62 – 63; Pl. 13: 1 – 6; Text-fig. 5.

pt. 1997 *Strachanognathus parvus* Rhodes – Armstrong, p. 789 – 791; Pl. 5: 1 – 3 (only); Text-fig. 8qt, ?ae, ?pf (only).

Remarks. – The few specimens conform with the description of Stouge (1984). Armstrong (1997) included "*Scolopodus*" *peselephantis* Lindström elements in the *Strachanognathus parvus* apparatus. This interpretation is not followed here because "*Scolopodus*" *peselephantis* appears for the first time long time before *Strachanognathus parvus* in the sections studied here (Tremadoc *versus* Llanvirn).

Occurrence. – Andersön-B, Steinsodden, Jøronlia, Høyberget, Engerdal. It is restricted to the *B. medius – H. holodentata* Zone within the Stein Formation, but is also recorded from the Elvdal Formation (*P. anserinus-A. inaequalis* Subzone).

Material. – 18 specimens.

Genus *Tetraprioniodus* Lindström, 1955

Type species. – *Tetraprioniodus robustus* Lindström, 1955

Revised diagnosis. – The apparatus of *Tetraprioniodus* comprises P, M and S elements. The apparatus reconstruction is not quite clear because of the small, badly preserved material at hand, but six element types have been noted. The undenticulated Pa element is distinct with its downward directed anterior process projecting 180° away from the erect cusp. Pb, Sb, and Sd elements have a strongly recurved cusp and bear denticulated posterior processes, whereas the Sa and M elements are adenticulate.

Remarks. – Bergström (1968, 1971) included *Tetraprioniodus robustus* s.f. in the *Prioniodus elegans* apparatus but the present material favours that *T. robustus sensu formo* instead makes the Sd element of a separate apparatus. Lindström (1971); Löfgren (1978) and Stouge & Bagnoli (1988) also questioned the relationship between *T. robustus sensu formo* and *P. elegans*. The multielement-genus *Tetraprioniodus* is presumably closely related to *Prioniodus* Pander 1856, but it has a very dissimilar P element, which is why it has been interpreted as being different from *Prioniodus*.

Tetraprioniodus robustus Lindström, 1955

Pl. 18: 2 – 10

Synonymy. –

Pa
1978 *Periodon?* prioniodontiform element A – Löfgren, p. 71; Pl. 10: 4, 6.

Pb
1974 *Gothodus microdentatus* n. sp. – Van Wamel, pp. 69 – 70; Pl. 5: 15a – b, ?14a – b.
1978 *Prioniodus?* (*Baltoniodus?*) sp. C – Löfgren, p. 89; Pl. 10: 13.

Sa
cf. 1955 *Trichonodella longa minor* n. var. – Lindström, p. 601; Pl. 6: 16 – 17 (only).
1978 *Periodon?* trichonodelliform element – Löfgren, p. 72; Pl. 10: 2A – B.

Sd
1955 *Tetraprioniodus robustus* n. sp. – Lindström, p. 597; Pl. 6: 13 – 15.
1978 *Prioniodus?* (*Baltoniodus?*) sp. B – Löfgren, p. 89; Pl. 10: 11A – B.
1978 *Periodon?* prioniodontiform element B – Löfgren, p. 71; Pl. 10: 5.

Multielement
1993a *Prioniodus? robustus* (Lindström) – Löfgren, Fig. 9v – w.
1994 *Prioniodus? robustus* (Lindström) – Löfgren, Fig. 8: 1 – 3, 8: 8 – 10, 16?

Description. –

Pa
The P element is "modified" pastinate with three adenticulate processes. The cusp is commonly erect but may be weakly proclined or reclined. The posterior and the postero-lateral processes are shorter than the anterior process and the cusp. A distinct basal sheath connects the processes. In mature specimens, however, the basal sheath is restricted to the proximal parts of the three processes. The anterior process forms an anticusp directed 180° downward from the cusp. The cusp and the anterior process are laterally compressed. The basal cavity is narrow and shallow along all processes.

Pb
The element is tertiopedate-dolabrate ramiform. The posterior process is denticulated, the anterior process adenticulate and the short, outer lateral, process is adenticulate. The denticles are variable in size but generally small. The cusp varies from suberect to strongly recurved. The element was described as *Gothodus microdentatus* n. sp. by van Wamel (1974, pp. 69 – 70).

M
The M element is geniculate coniform and adenticulate. The antero-basal extension is triangular. The inner side of the element is thickened near the cusp-base junction. The cusp is twisted and carinate. The

angle between the cusp and the upper margin is about 35°.

Sa
The element is alate ramiform. Serrations may occur on the upper margin of the processes but denticles in the proper sense have not been observed. The angle between the lateral processes is usually about 40° in an upper view, but may be as high as 50°.

Sb
The element is tertiopedate with a denticulate, posterior process, a adenticulate, anterior process and a adenticulate, lateral process. The lateral and anterior processes deflect backwards. A distinct, sharp-edged costa is situated on the outer surface of the cusp. It continues across the base and on the outer lateral process. The lateral process is slightly shorter than the anterior process. The cusp is recurved.

Sd
The quadriramate Sd element was fully described as the form-species *Tetraprioniodus robustus* n. sp. by Lindström (1955, p. 597).

Remarks. – Löfgren (1993a, 1994) revised the concept of *T. robustus* (*Prioniodus? robustus* of Löfgren) and her apparatus reconstruction agrees closely with the one presented here. However, the P (prioniodonti-form) element of Löfgren (1993, Fig. 9w) is interpreted here as the Pb element. *T. robustus* is similar to *Prioniodus gilberti* Stouge & Bagnoli 1988 with respect to the Sa and Sb elements. *T. robustus*, however, has an apparatus of which Sd elements are denticulated. In comparison *P. gilberti* is characterised by an adenti-culate Sd element. It was suggested already by van Wamel (1974) that the form-species *Gothodus micro-dentatus* van Wamel (the Pb element herein) likely was part of a multielement-species.

Occurrence. – Andersön-A, -B. *P. proteus* Zone.

Material. – 23 Pa, 9 Pb, 4 M, 5 Sa, 1 Sb, 12 Sd.

Genus *Trapezognathus* Lindström, 1955

Type species. – *Trapezognathus quadrangulum* Lind-ström, 1955

Remarks. – The multielement genus *Trapezognathus* was reconstructed and described by Stouge & Bagnoli (1990), who established the septimembrate apparatus which comprises Pa, Pb, M, Sa, Sb, Sc and Sd elements.

Trapezognathus quadrangulum Lindström, 1955

Pl. 19: 12 – 17

Synonymy. –

pt. 1955a *Trapezognathus quadrangulum* n. sp. – Lind-ström, p. 598; Pl. 5: 38, 39 (only).
1990 *Trapezognathus quadrangulum* Lindström – Stouge & Bagnoli, pp. 26 – 27; Pl. 10: 1 – 5, 7 – 10 (*cum. syn.*).
1993b *Lenodus?* sp. – Löfgren, Fig. 6I.
1994 *Lenodus?* sp. A – Löfgren, Fig. 8: 24 – 27.
1997 *Trapezognathus quadrangulum* Lindström – Bagnoli & Stouge, p. 160; Pl. 8: 1 – 8.

Remarks. – *Trapezognathus quadrangulum* was described by Stouge & Bagnoli (1990). The apparatus is similar to *Baltoniodus triangularis* Lindström but differs by the presence of a strongly recurved cusp in Sa and Sd elements, an angle of 90° or more between the upper margin and the cusp in M elements, and a different denticulation on the lateral processes in Sa elements.

The Sa, Sd and M elements are similar to the corresponding elements in *Lenodus* Sergeeva, and *Trape-zognathus* is likely the ancestor of this species (Stouge & Bagnoli 1990, p. 17).

Occurrence. – Herram, Stein. *M. flabellum* – *D. forceps* Zone – *P. zgierzensis* Zone (basal part).

Material. – 6 Pa, 6 Pb, 5 M, 7 Sa, 4 Sb, 3 Sc, 6 Sd.

Genus *Triangulodus* van Wamel, 1974

Type species. – *Paltodus volchovensis* Sergeeva, 1963

Remarks. – The seximembrate apparatus was recon-structed by van Wamel (1974). It comprises one nongeniculate coniform (scandodontiform) P element, four nongeniculate, costate coniform S elements and one geniculate coniform M element. *Triangulodus brevibasis* (Sergeeva) has earlier been referred to the genus *Trigonodus* Crespin (e.g. Stouge & Bagnoli 1990; Rasmus-sen 1991; Löfgren 1993b) but as *Trigonodus* already was occupied when Crespin (1943) proposed the name, the name is invalid (Nicoll in Stait & Druce 1993).

Triangulodus amabilis n. sp.

Pl. 19: 1 – 7

Derivation of name. – Amabilis (Latin); lovely.

Type locality. – Steinsodden, Moelv, Norway.

Type stratum. – 0.1 m above the base of the Stein Formation at Steinsodden (sample 69687), *M. flabellum* – *D. forceps* Zone.

Holotype. – Repository PMO 165.485. Sd element (Plate 19: 7).

Diagnosis. – The S elements are characterised by relatively long process-like extensions of the costae. The elements are commonly hyaline, but may be partly albid. The asymmetrical, quadriramate Sd element is laterally compressed, with two anterior and two posterior costae. The outer surface of the Sd element is clearly smaller than the inner.

Description. –

P

The P element is nongeniculate coniform with a proclined or suberect cusp. The element is inwardly bowed and carries an indistinct carina on the inner surface of the cusp. Costae are situated on the anterior and posterior edges of the cusp and on the upper margin. The angle between the upper margin and the cusp is 100° – 120°. The basal margin is sinuous. The basal cavity is triangular in lateral view.

M

The M element is geniculate coniform with a reclined or recurved cusp. A faint carina is placed on the inner surface of the cusp. The basal margin is weakly convex. The downward directed extension of the anterior costa forms a characteristic antero-basal corner making an acute angle.

S elements, general remarks

Costae extend over the basal margin and make short, adenticulated processes.

Sa

The element is alate coniform with two lateral, symmetrically arranged costae situated at both sides of the posterior costa. In cross section, the element is triangular with a weakly concave anterior margin and an acute-angled posterior/upper margin. The basal margin is concave between the processes. The cusp is proclined.

Sb

The element is tertiopedate coniform with one anterior, one posterior and one outer-lateral costa. The basal margin is strongly concave between the processes. Upper margin is weakly convex. The cusp is proclined. The basal cavity is triangular in side view.

Sc

The element is dolabrate coniform. The cusp and base are laterally compressed. The suberect cusp is keeled. The

anterior keel continues across the base and forms a characteristic, process-like, flattened, antero-basal extension. The basal margin is concave.

Sd

Sd element is quadriramate coniform with four costae arranged "two anterior – two posterior". The element is asymmetrical around the vertical, anterior – posterior mirror plane, meaning that the outer surface is clearly smaller than the inner. Both surfaces are almost planar. The basal margin is slightly concave, while the upper margin is weakly convex.

Remarks. – *Triangulodus amabilis* may be distinguished from *T. brevibasis* by the distinct Sd element and the generally longer processes of the S elements. It differs from *Triangulodus? alatus* (Dzik, 1976) by the shorter processes. Dzik (1976, Fig. 20f – k) did not illustrate the Sd or M elements of *T.? alatus*.

The Sd element resembles the quadriramate element of *Tropodus comptus australis* (e.g. Landing 1976, Pl. 4: 22 – 23 [only]) but is distinguished from this by the wider angle between the cusp and the upper margin of the base, and by having a convex rather than concave upper margin. *T. amabilis* is somewhat similar to the stratigraphically younger *Triangulodus maocaopus* Zhang, but is distinguished by its different M, Sc and Sd elements. *T. amabilis* may be the ancestor of *T. maocaopus*.

Occurrence. – Steinsodden. *M. flabellum* – *D. forceps* Zone – *P. zgierzensis* Zone.

Material. – 15 P, 6M, 6 Sa, 10 Sb, 5 Sc, 4 Sd.

Triangulodus brevibasis (Sergeeva, 1963)

Pl. 19: 8 – 1.

Synonymy. –

M

1963 *Oistodus brevibasis* n. sp. – Sergeeva, p. 95; Pl. 7: 4 – 5; Text-fig. 2.

Multielement

1971 *Scandodus brevibasis* (Sergeeva) – Lindström, p. 39; Pl. 1: 24 – 27; Text-fig. 3.

1974 *Triangulodus brevibasis* (Sergeeva) – van Wamel, p. 96; Pl. 5: 1 – 7.

1978 *Scandodus brevibasis* (Sergeeva) – Löfgren, p. 104; Pl. 1: 30 – 35 (*cum. syn.*).

1990 *Trigonodus brevibasis* (Sergeeva) – Stouge & Bagnoli, p. 28; Pl. 10: 18 – 26.

1991 *Trigonodus brevibasis* (Sergeeva) – Rasmussen, p. 285; Fig. 8R – U.

cf. 1993   *Triangulodus* cf. *brevibasis* (Sergeeva) – Stait & Druce, p. 315; Fig. 20J – L, N – O; Fig. 21K; Text-fig. 14D – F.

1994   *Triangulodus brevibasis* (Sergeeva) – Löfgren, Fig. 25 – 27.

1995a   *Triangulodus brevibasis* (Sergeeva) – Lehnert, p. 129; Pl. 8: 15.

1995b   *Triangulodus brevibasis* (Sergeeva) – Löfgren, Fig. 9ah – am.

1998b   *Triangulodus brevibasis* (Sergeeva) – Zhang, pp. 93 – 94; Pl. 18: 1 – 2.

Remarks. – The apparatus of *Triangulodus brevibasis* was reconstructed and described by van Wamel (1974).

The general morphology of *Triangulodus larapintensis* (Crespin 1943, *sensu* Cooper 1981) resembles *T. brevibasis*, but the latter species is characterised by its distinctive basal cavity (Stait & Druce 1993).

Occurrence. – Andersön-A, -B. *M. flabellum – D. forceps* Zone – *P. rectus – M. parva* Zone.

Material. – 12 P, 8 M, 2 Sa, 6 Sb, 2 Sd.

Gen. et sp. indet. A

Pl. 18: 11

Synonymy. –

Pa?
? 1955   *Ambalodus* n. sp. – Lindström, p. 550; Pl. 6: 45, 46.
1988   Platform element A – Bagnoli & Stouge, p. 216; Pl. 41: 14.

Pb?
? 1974   *Fryxellodontus? corbatoi* n. sp. – Serpagli, pp. 47 – 48; Pl. 10: 1a – 6c, Pl. 22: 1 – 5.
? 1988   *Polonodus? corbatoi* Serpagli – Stouge & Bagnoli, p. 131; Pl. 10: 1 – 5 (*cum. syn.*).

Description. –

?Pa
The element is stelliplanate pectiniform with one anterior, one posterior and one lateral process. All processes carry a median row of nodes. The basal cavity is wide and deep.

?Pb
The element is an alate pectiniform with an anterior, cusp-like, rounded apex. The one posterior and two lateral processes (or ridges) carry nodes and are connected by walls, which may be interpreted as a large basal sheath.

Remarks. – The Pa? element resembles *Ambalodus* n. sp.

sensu Lindström (1955a) but because the former element is badly preserved and lacks the distal part of the processes, the relationship with this species is questionable.

The scale-like microsculptured surface of *Polonodus? corbatoi*, illustrated by Stouge & Bagnoli (1988), was not observed in the Pb? specimen at hand.

Occurrence. – Steinsodden, Andersön-B. *P. proteus* Zone.

Material. – 2 Pa?, 1 Pb?.

Gen. et sp. indet. B

Pl. 18: 12

Description. – The element is a weakly inwardly bowed, nongeniculate coniform. Sharp edges are formed on the anterior and posterior margins. The cusp is proclined. The angle between the upper margin and the anterior margin of the biconvex base approximates 60° in lateral view. The basal cavity is triangular with the apex near the anterior margin of the element. The basal cavity extends about ⅓ of the total length.

Remarks. – The element belongs possibly to *Costiconus* n. gen., but the material at hand is too small to prove this.

Occurrence. – Andersön-B, Steinsodden. *P. graeai* Zone (upper part) – *E. suecicus* Zone.

Material. – 10 scandodontiform specimens.

Gen. et sp. indet. C

Pl. 18: 14 – 15

Synonymy. –

1978   Gen. et sp. nov. – Löfgren, pp. 117 – 118; Fig. 34.
pt. 1987   *Stolodus stola stola* (Lindström) – Olgun, p. 55; Pl. 7ZZ (only).

Description. –

P
The element is tricostate pectiniform with one anterior, one posterior and one outer-lateral costa. No denticles have been observed. The angle between the anterior and posterior processes is about 55°. The walls are thin. The cusp is proclined and short. The basal cavity is deep and extends to the tip of the cusp.

Sd

The element is an adenticulate, quadricostate ramiform carrying two anterior and two posterior costae. The walls are thin and slightly convex between the costae. The angle between the two posterior processes is 35° in upper view. The cusp is proclined and short. The basal cavity is deep and extends to the tip of the cusp.

Remarks. – Gen. et sp. indet. C share some characteristics with *Stolodus stola* Lindström and *Ansella jemtlandica* (Löfgren) by having a very deep basal cavity, thin walls and a proclined cusp. The Sd element, however, has slightly convex surfaces and a more robust outline than *Stolodus stola*, while the *Ansella* apparatus lacks the quadricostate element. Löfgren (1978) recorded this species from levels correlating with the *P. rectus – M. parva* and *B. norrlandicus – D. stougei* zones in the autochthonous succession in Jämtland.

Occurrence. – Steinsodden. *P. rectus – M. parva Zone.*

Material. – 2 P? and 1 Sd element.

Gen. et sp. indet. D

Pl. 18: 13

Description. –

P, general remarks
The elements are pastinate pectiniform with one anterior, one posterior and on outer-lateral process, all of which are denticulated. The lateral process carries an anterior denticle row. The denticle rows of the anterior and the lateral processes are joined in the denticle just in front of the cusp. The suberect cusp is short and robust. The basal cavity is wide and extends to the tip of the processes.

Pa
The angle between the anterior and posterior processes is about 120° in a lateral view, while the angle between the lateral and posterior processes is about 80°. A prominent, thickened ledge is developed on the outer margin of the posterior and lateral processes.

Pb
The angle between the anterior and posterior processes approximates 80° in lateral view, while the angle between the lateral and posterior processes is about 45°.

Remarks. – Gen. et sp. indet. D show some resemblance with the stratigraphically younger, Middle to Late Ordovician, *Rhodesognathus elegans* Rhodes, but the former is characterised by smaller denticles, a shorter cusp and wider processes. It is possible that Gen. et sp. indet. D belongs to *Lenodus* Sergeeva *sensu* Stouge & Bagnoli 1988 or *Polonodus* Dzik but more material is needed to confirm this.

Occurrence. – Andersön-B. *B. medius – H. holodentata* Zone (basal part).

Material. – 2 Pa, 3 Pb.

Acknowledgements. – My thanks are due foremost to Svend Stouge (Geological Survey of Denmark and Greenland) and Valdemar Poulsen and Eckart Håkansson (University of Copenhagen) for their advice, motivation and help throughout the Ph.D. study on which this paper is based (Rasmussen 1994). I am indebted to Stig Bergström (Ohio State University) for his careful review of the present work, and to Anita Löfgren (University of Lund) for many helpful suggestions and comments on an earlier version of the manuscript. The project was initiated at the Geological Institute, University of Copenhagen. The Ph.D. study was financed by a faculty grant (*kandidatstipendium*) from the University of Copenhagen, which is gratefully acknowledged. A grant from *Nordisk Ministerråd* (31043.31.156/90) paid for a four-month visit at the Palaeontological Museum in Oslo and is greatly acknowledged. The study was also supported by the Danish Natural Science Research Council (grant no. 28808 to Svend Stouge). David Bruton (University of Oslo) deserves special thanks for his hospitality and advice during my stay in Oslo. David Harper and Arne Thorshøj Nielsen (Geological Museum, University of Copenhagen) are sincerely thanked for fruitful discussion on Ordovician faunas and stratigraphy and DH also for reading many parts of the text from a linguistic point of view. Field assistance by Tage Rasmussen, SEM assistance by Jørgen Fuglsang, graphic help from Henrik Egelund and Stefan Sølberg, photographs by Jan Aagaard and atomic-absorption spectro-photometer measurements by Birthe Damgaard are gratefully acknowledged. I thank Maria Hallquist, Mads Willumsen, Uffe Juul, Karen Daugbjerg, Peter Hegel and many other friends and colleagues from the Geological Institute for technical support during the final stage of the present work. Austin Boyd and David Bridgewater read an earlier version of the manuscript from a linguistic point of view. The Geological Survey of Denmark and Greenland is thanked for permission to publish the manuscript. Publication subsidies from Statens Naturvidenskabelige Forskningsråd (Danish Natural Science Research Council) are deeply appreciated (grant no. 51-00-0948 to Hans Jørgen Hansen).

References

Albanesi, G.L., Hünicken, M.A., & Ortega, G. 1995: Review of Ordovician conodont-graptolite biostratigraphy of the Argentine Precordillera. *In* Cooper, J.D., Droser, M.L. & Finney, S.C. (eds.): Ordovician Odyssey: Short Papers for the Seventh International Symposium on the Ordovician System. *SEPM, Pacific Section, Book 77*, 31–35.

Albanesi, G.L., & Barnes, C.R. 1996: The origin of the Middle Ordovician conodont *Paroistodus horridus* in the Argentine Precordillera. *Sixth European Conodont Symposium (ECOS VI), Abstracts*, p. 1. Instytut Paleobiologii PAN, Warszawa.

Aldridge, R.J. (ed.) 1987: Palaeobiology of conodonts. *British Micropalaeontological Society, Ellis Harwood Limited,* Chichester. 180 pp.

Aldridge, R.J., Briggs, D.E.G., Smith, M.P., Clarkson, E.N.K., & Clark, N.D.L. 1993: The anatomy of conodonts. *Philosophical Transactions of the Royal Society of London B 340,* 405–421.

Aldridge, R.J. & Smith, M. P. 1993: Conodonta. *In* Benton, M.J. (ed.): *The Fossil Record 2,* 563–572. Chapman & Hall, London.

An, T. 1981: Recent progress in Cambrian and Ordovician conodont biostratigraphy of China. *In* Teichert, C., Liu, Lu. & Chen Peiji (eds.): *Paleontology in China. Geological Society of America Special Paper 187,* 209–224.

An, T., Du, G. & Gao, Q. 1985: [Ordovician conodonts from Hubei, China]. 64 pp. Bejing Geological Publishing House. (In Chinese with English summary.)

Armstrong, H.A. 1997: Conodonts from the Ordovician Shinnel Formation, Southern Uplands, Scotland. *Palaeontology 40,* 763–797.

Asklund, B. 1933: Vemdalskvartsitens ålder. *Sveriges Geologiska Undersökning, Serie C, Årsbok 27,* 1–56.

Asklund, B. 1960: The geology of the Caledonian mountain chain and of adjacent areas in Sweden. *In* Magnusson, N.H., Thorslund, P., Brotzen, F., Asklund, B. & Kulling, O. (eds.): *Description to accompany the map of the Pre-Quaternary rocks of Sweden. Sveriges Geologiska Undersökning Ba 16,* 126–149.

Bagnoli, G., Barnes, C.R. & Stevens, R.K. 1987: Lower Ordovician (Tremadocian) conodonts from Broom Point and Green Point, western Newfoundland. *Bollettino della Societa Paleontologica Italiana 25,* 145–158.

Bagnoli, G. & Stouge, S. 1997: Lower Ordovician (Billingenian - Kunda) conodont zonation and provinces based on sections from Horns Udde, north Öland, Sweden. *Bollettino della Societa Paleontologica Italiana 35,* 109–163.

Bagnoli, G., Stouge, S.S. & Tongiorgi, M. 1988: Acritarchs and conodonts from the Cambro-Ordovician Furuhäll (Köpings-klint) section (Öland, Sweden). *Rivista Italiana de Paleontologica e Stratigrafia 94,* 163–248.

Barnes, C.R. & Poplawski, M.L.S. 1973: Lower and Middle Ordovician conodonts from the Mystic Formation, Quebec, Canada. *Journal of Paleontology 47,* 760–790.

Barrick, J.E. 1977: Multielement simple-cone conodonts from the Clarita Formation (Silurian), Arbuckle Mountains, Oklahoma. *Geologica et Palaeontologica 11,* 47–68.

Bauer, J.A. 1989: Conodonts biostratigraphy and palaeoecology of Middle Ordovician rocks in Eastern Oklahoma. *Journal of Palaeontology 63(1),* 92–107.

Bednarczyk, W.S. 1998: Ordovician conodont biostratigraphy of the Polish part of the Baltic Syneclise. *In* Szaniawski, H. (ed.): Proceedings of the Sixth European Conodont Symposium (ECOS VI). *Palaeontologica Polonica 58,* 107–121.

Bengtson, P. 1988: Open nomenclature. *Palaeontology 31(1),* 223–227.

Bergström, S.M. 1962: Conodonts from the Ludibundus Limestone (Middle Ordovician) of the Tvären area (S.E. Sweden). *Arkiv för Mineralogi och Geologi 3(1),* 1–61.

Bergström, S.M. 1971: Conodont biostratigraphy of the Middle and Upper Ordovician of Europe and eastern North America. *In* Sweet, W.C., Bergström S.M. (eds.): *Symposium on Conodont Biostratigraphy. Geological Society of America Memoir 127,* 83–161.

Bergström, S.M. 1973: Biostratigraphy and facies relations in the Middle Ordovician of easternmost Tennessee. *American Journal of Science 273-A,* 261–293.

Bergström, S.M. 1978: Middle and Upper Ordovician conodont and graptolite biostratigraphy of the Marathon, Texas graptolite zone reference standard. *Palaeontology 21(4),* 723–758.

Bergström, S.M. 1979: Whiterockian (Ordovician) conodonts from the Hølonda Limestone of the Trondheim Region, Norwegian Caledonides. *Norsk Geologisk Tidsskrift 59,* 295–307.

Bergström, S.M. 1983: Biostratigraphy, evolutionary relationships, and stratigraphic significance of Ordovician platform conodonts. *In* Martinson, A. & Bengtson, S. (eds.): *Taxonomy, Ecology and Identity of Conodonts. Proceedings of the Third European Conodont Symposium (ECOS III) in Lund, 30th August to 1st September. Fossils and Strata 15,* 35–58.

Bergström, S.M. 1986: Biostratigraphic integration of Ordovician graptolite and conodont zones - a regional review. *In* Hughes, C.P. & Rickards, R.B. (eds.): *Palaeoecology and Biostratigraphy of Graptolites. Geological Society of London Special Publications 20,* 61–78.

Bergström, S.M. 1988: On Pander's Ordovician conodonts: distribution and significance of the *Prioniodus elegans* fauna in Baltoscandia. *Senckenbergiana lethaea 69,* 217–251.

Bergström, S.M. 1990: Biostratigraphic and biogeographic significance of Middle and Upper Ordovician conodonts in the Girvan succession, south-west Scotland. *Courier Forschungs-Institut Senckenberg 118,* 1–43.

Bergström, S.M. 1997: Conodonts of Laurentian faunal affinities from the Middle Ordovician Svartsætra limestone in the Trondheim Region, Central Norwegian Caledonides. *Norges geologiske undersøkelse Bulletin 432,* 59–69.

Bergström, S.M., Epstein, A.G. & Epstein, J.B. 1972: Early Ordovician North Atlantic province conodonts in eastern Pennsylvania. *United States Geological Survey Professional Paper 800D,* D37-D44.

Bergström, S.M. & Orchard, M.J. 1985: Conodonts from the Cambrian and Ordovician Systems. *In* Higgins, A.C. & Austin, R.L. (eds.): *A Stratigraphical Index of Conodonts. British Micropalaeontological Society Series,* 32–67. Ellis Horwood Limited, Chichester.

Bergström, S.M. & Sweet, W.C. 1966: Conodonts from the Lexington Limestone (Middle Ordovician) of Kentucky and its lateral equivalents in Ohio and Indiana. *Bulletin of American Paleontology 50,* 269–441.

Bergström, S.M., Riva, J. & Kay, M. 1974: Significance of conodonts, graptolites, and shelly faunas from the Ordovician of western and north-central Newfoundland. *Canadian Journal of Earth Sciences 11,* 1625–1660.

Bergström, S.M., Rhodes, F.H.T. & Lindström, M. 1987: Conodont biostratigraphy of the Llanvirn - Llandeilo and Llandeilo - Caradoc Series boundaries in the Ordovician System of Wales and the Welsh Borderland. *In* Austin, R.L. (ed.): *Conodonts: Investigative Techniques and Applications. British Micropalaeontological Society Series,* 294–315. Ellis Horwood Limited, Chichester.

Berry, W.B.N. 1964: The Middle Ordovician of the Oslo Region, Norway. 16. Graptolites of the Ogygiocaris Series. *Norsk Geologisk Tidsskrift 44,* 61–170.

Bjørlykke, A. 1973: DOKKA, berggrunnsgeologisk kart 1816 IV-M. 1:50.000 *Norges Geologiske Undersøkelse.*

Bjørlykke, A. 1979: Gjøvik og Dokka. Description of the geological maps 1816 I and 1816 IV-1:50000 *Norges Geologiske Undersøkelse 344,* 1–48.

Bjørlykke, A. & Skålvoll H. 1979: BRUFLAT, berggrunnsgeologisk kart 1716 I-M. 1:50.000 *Norges Geologiske Undersøkelse.*

Bjørlykke, K.O. 1905: Det centrale Norges Fjeldbygning. *Norges Geologiske Undersøkelse 39,* 610 pp.

Bockelie, J.F. 1978: The Oslo Region during the Early Palaeozoic. *In* Ramberg, I.B. & Neumann, E.R. (eds.): *Tectonics and Geophysics of Continental Rifts*, 195–202. D. Reidel, Dordrecht.

Bockelie, J.F. & Nystuen, J.P. 1985: The southeastern part of the Scandinavian Caledonides. *In* Gee, D.G. & Sturt, B.A. (eds.): *The Caledonide Orogen - Scandinavia and Related Areas*, 69–88. John Wiley & Sons Limited, Chichester.

Bradshaw, L.E. 1969: Conodonts from the Fort Peña Formation (Middle Ordovician), Marathon Basin, Texas. *Journal of Palaeontology 43(5)*, 1137–1168.

Branson, E.B. & Branson, C.C. 1947: Lower Silurian conodonts from Kentucky. *Journal of Paleontology 21*, 549–556.

Branson, E.B. & Mehl, M.G. 1933: Conodont studies. *University of Missouri Studies 8*, 1–349.

Briggs, D.E.G., Clarkson, E.N.K. & Aldridge, R.J. 1983: The conodont animal. *Lethaia 16*, 1–14.

Bruton, D.L. & Harper, D.A.T. 1985: Early Ordovician (Arenig-Llanvirn) faunas from oceanic islands in the Appalachian-Caledonide orogen. *In* Gee, D.G. & Sturt, B.A. (eds.): *The Caledonide Orogen - Scandinavia and Related Areas*, 359–368. John Wiley & Sons Limited, Chichester.

Bruton, D.L. & Harper, D.A.T. 1988: Arenig - Llandovery stratigraphy and faunas across the Scandinavian Caledonides. *In* Harris, A.L. & Fettes, D.J. (eds.): *The Caledonian-Appalachian Orogen. Geological Society Special Publications 38*, 247–268.

Bruton, D.L., Harper, D.A.T. & Repetski, J.E. 1989: Stratigraphy and faunas of the Parautochthon and Lower Allochthon of southern Norway. *In* Gayer, R.A. (ed.): *The Caledonide Geology of Scandinavia*, 231–241. Graham & Trotman.

Cooper, B.J. 1976: Multielement conodonts from the St. Clair Limestone (Silurian) of southern Illinois. *Journal of Paleontology 50*, 205–217.

Cooper, B.J. 1981: Early Ordovician conodonts from the Horn Valley Siltstone, central Australia. *Palaeontology 24*, 147–183.

Crespin, I. 1943: Conodonts from the Waterhouse Range, Central Australia. *Transactions of the Royal Society of South Australia 67*, 231–233.

Dunham, R.J. 1962: Classification of carbonate rocks according to depositional texture. *In* Ham, W.E. (ed.): *Classification of Carbonate Rocks. American Association of Petroleum Geologists 1*, 108–121.

Dzik, J. 1976: Remarks on the evolution of Ordovician conodonts. *Acta Palaeontologica Polonica 21*, 395–455.

Dzik, J. 1978: Conodont biostratigraphy and Paleogeographical relations of the Ordovician Mojcza Limestone (Holy Cross Mts, Poland). *Acta Palaeontologica Polonica 23(1)*, 51–69.

Dzik, J. 1983: Early Ordovician conodonts from the Barrandian and Bohemian-Baltic faunal relationships. *Acta Palaeontologica Polonica 28*, 327–368.

Dzik, J. 1989: Conodont evolution in high latitudes of the Ordovician. *In* Ziegler, W. (ed.): *Papers on Ordovician to Triassic Conodonts. 1st International Senckenberg Conference and 5th European Conodont Symposium (ECOS V) Contributions III*, 1–28.

Dzik, J. 1990: Conodont evolution in high latitudes of the Ordovician. *Courier Forschungsinstitut Senckenberg 117*, 1–28.

Dzik, J. 1991: Evolution of oral apparatuses in the conodont chordates. *Acta Palaeontologica Polonica 36*, 265–323.

Dzik, J. 1994: Conodonts of the Mojcza Limestone. *In* Dzik, D., Olempska, E. & Pisera, A.: *Ordovician Carbonate Platform Ecosystem of the Holy Cross Mountains. Palaeontologica Polonica 53*, 43–128.

Ethington, R.L. 1959: Conodonts of the Ordovician Galena Formation. *Journal of Paleontology 55*, 239–247.

Ethington, R.L. & Clark, D.L. 1971: Lower Ordovician conodonts in North America. *In* Sweet, W.C. & Bergström, S.M. (eds.): *Symposium on Conodont Biostratigraphy. Geological Society of America, Memoir 127*, 63–82.

Ethington, R.L. & Clark, D.L. 1982: Lower and Middle Ordovician conodonts from the Ibex area, western Millard County, Utah. *Brigham Young University Geological Studies 28(2)*, 1–155.

Ethington, R.L. & Schumacher, D. 1969: Conodonts from the Copenhagen Formation (Middle Ordovician) in central Nevada. *Journal of Paleontology 43*, 440–484.

Finney, S.C. & Ethington, R.L. 1992: Whiterockian graptolites and conodonts from the Vinini Formation, Nevada: Biostratigraphic implications. *In* Webby, B.D. & Laurie, J.R. (eds.): *Global Perspectives on Ordovician Geology*, 153–169. A.A. Balkema, Brookfield.

Fåhræus, L.E. 1966: Lower Viruan (Middle Ordovician) conodonts from the Gullhögen quarry, southern central Sweden. *Sveriges Geologiska Undersökning C610*, 1–40.

Fåhræus, L.E. 1970: Conodont-based correlations of Lower and Middle Ordovician strata in Western Newfoundland. *Geological Society of America Bulletin 81*, 2061–2076.

Fåhræus, L.E. 1982: Recognition and redescription of PANDER's (1856) Scolopodus (form-) species - Constituents of multi-element taxa (Conodontophorida, Ordovician). *Geologica et Palaeontologica 13*, 19–28.

Fåhræus, L.E. 1984: A critical look at the Treatise family-group classification of Conodonta: an exercise in eclectism. *Lethaia 17*, 293–305.

Fåhræus, L.E. & Hunter, D.R. 1981: Paleoecology of selected conodontophorid species from the Cobbs Arm Formation (Middle Ordovician), New World Island, north-central Newfoundland. *Canadian Journal of Earth Sciences 18*, 1653–1665.

Fåhræus, L.E. & Hunter, D.R. 1985: Simple-cone conodont taxa from the Cobbs Arm Limestone (Middle Ordovician), New World Island, Newfoundland. *Canadian Journal of Earth Sciences 22*, 1171–1182.

Fåhræus, L.E. & Hunter, D.R. 1985: The curvature-transition series: integral part of some simple-cone apparatuses (Panderodontacea, Distacodontacea, Conodontata). *Acta Palaeontologica Polonica 30*, 177–189.

Fåhræus, L.E. & Nowlan, G.S. 1978: Franconian (Late Cambrian) to early Champlainian (Middle Ordovician) conodonts from the Cow Head Group, Western Newfoundland. *Journal of Paleontology 52*, 444–471.

Gedik, I. 1977: Orta Toroslar'da konodont biyostratigrafisi (Conodont biostratigraphy in the Middle Taurus). *Türkiye Jeoloji Kurumu Buëlteni 20*, 35–48.

Gee, D.G. 1975: A tectonic model for the central part of the Scandinavian Caledonides. *American Journal of Science 275-A*, 468–515.

Gee, D.G., Guezou, J.-C., Roberts, D. & Wolff, F.C. 1985: The central-southern part of the Scandinavian Caledonides. *In* Gee, D.G. & Sturt, B.A. (eds.): *The Caledonide Orogen-Scandinavia and Related Areas*, 109–133. John Wiley & Sons, Chichester.

Graves, R.W. & Ellison, S. 1941: Ordovician conodonts of the Marathon Basin, Texas. *University of Missouri, School of Mines & Metallurgy Bulletin, Technichal Series 14(2)*, 1–26.

Hadding, A.R. 1912: Några iakttagelser från Jämtlands Ordovicium. *Geologiska Föreningens i Stockholm Förhandlingar 34*, 589–602.

Hadding, A.R. 1913: Undre dicellograptusskiffern i Skåne jämte några därmed ekvivalenta bildningar. *Lunds Universitets Årsskrift., N. F., Avd 2, 9(15)*, 90 pp.

Hamar, G. 1964: Conodonts of the lower Middle Ordovician of Ringerike. *Norsk Geologisk Tidsskrift 44*, 243–292.

Hamar, G. 1966: Preliminary report on conodonts from the Oslo-Asker and Ringerike districts. *Norsk Geologisk Tidsskrift 46*, 27–83.

Harper, D.A.T. & Rasmussen, J.A. 1997 (for 1996): Phosphatic microbrachiopod biofacies in the Lower Allochthon of the Scandinavian Caledonides. *In* Stouge, S. (ed.): WOGOGOB-94 Symposium, Bornholm-94. *Danmarks og Grønlands Geologiske Undersøgelse, Rapport 1996/98*, 33–42.

Hass, W.H. 1962: Conodonts. *In* Moore, R.C. (ed.): *Treatise on Invertebrate Paleontology, Part W, Miscellanea*, W3–W69. *Geological Society of America and University of Kansas*, Boulder.

Heath, R.A. & Owen, A.W. 1991: Stratigraphy and biota across the Ordovician-Silurian boundary in Hadeland, Norway. *Norsk Geologisk Tidsskrift 71*, 91–106.

Henningsmoen, G. 1948: The Tretaspis Series of the Kullatorp Core. *In* Wærn B., Thorslund, P. & Henningsmoen, G. Deep boring through Ordovician and Silurian Strata at Kinnekulle, Vestergötland. *Bulletin of the Geological Institution of the University of Uppsala 32*, 374–432.

Henningsmoen, G. 1979: Østerdal-Trysil-Engerdal-traktenes kambro-ordoviciske fossiler (annen og sidste del). *Fossil-Nytt 1979*, 3–22.

Hibbard, J.P., Stouge, S. & Skevington, D. 1977: Fossils from the Dunnage Mélange, north-central Newfoundland. *Canadian Journal of Earth Science 14*, 1176–1178.

Hinde, G.J. 1879: On conodonts from the Chazy and Cincinnati group of the Cambro-Silurian and from the Hamilton and Genesee-shale divisions of the Devonian in Canada and the United States. *Quarterly Journal of the Geological Society, London 35*, 351–369.

Holmsen, G. 1937: Søndre Femund: Beskrivelse til det geologiske rektangelkart. *Norges Geologiske Undersøkelse 148*, 41 pp.

Holtedahl, O. 1909: Studien über die Etage 4 des norwegischen Silursystems beim Mjösen. *Videnskapernes Selskab Skrifter, M.-nat. Kl.I, 1(7)*, 1–76.

Holtedahl, O. 1920: Kalksten og dolomit i de ostlandske dalfører. *Norges Geologiske Undersøkelse, Årbok 1920, 1*, 31 pp.

Holtedahl, O. 1921: Engerdalen, fjeldbygningen inden rektangelkartet Engerdalens omraade. *Norges Geologiske Undersøkelse 89*, 74 pp.

Hossack, J.R., Garton, M.R. & Nickelsen, R.P. 1985: The geological section from the foreland up to the Jotun thrust sheet in the Valdres area, south Norway. *In* Gee, D.G. & Sturt, B.A. (eds.): *The Caledonide Orogen-Scandinavia and Related Areas*, 443–456. John Wiley & Sons Limited, Chichester.

Hünicken, M.A. & Ortega, G.C. 1987: Lower Llanvirn-Lower Caradoc (Ordovician) conodonts and graptolites from the Argentine Central Precordillera. *In* Austin, R.L. (ed.): *Conodonts: Investigative Techniques and Applications*. British Micropalaeontological Society Series, 136–145. Ellis Horwood Limited, Chichester.

Högbom, A.G. 1891: Kvartsit-Sparagmitområdet i Sveriges sydliga fjelltrakter. *Sveriges Geologiska Undersökning, Ser. C, 116*, 20 pp.

Högbom, A.G. 1920: Geologisk beskrivning över Jämtlands län. *Sveriges Geologiska Undersökning, Ser. C, 140*, 138 pp.

Høy, T. & Bjørlykke, A. (1980). HAMAR, berggrunnskart 1916 IV-M. 1:50000.

Jaanusson, V. 1972: Constituent analysis of an Ordovician limestone from Sweden. *Lethaia 5*, 217–237.

Jaanusson, V. 1976: Faunal dynamics in the Middle Ordovician (Viruan) of Balto-Scandia. *In* Basset, M. (ed.): *The Ordovician System. Proceedings of the Palaeontological Association, Symposium, Birmingham, September 1974*, 301–326, University of Wales Press and National Museum of Wales, Cardiff.

Jaanusson, V. 1982: Introduction to the Ordovician of Sweden. *In* Bruton, D.L. & Williams, S.H. (eds.): *Field Excursion Guide. IVth International Symposium on the Ordovician System. Palaeontological Contributions from the University of Oslo 279*, 1–9.

Karis, L. 1982: The sequence in the Lower Allochthon of Jämtland. *In* Bruton, D.L. & Williams, S.H. (eds.): *Field Excursion Guide. IVth International Symposium on the Ordovician System. Palaeontological Contributions from the University of Oslo 279*, 55–63.

Kjerulf, T. 1863: Om et fund af fossiler ved Högberget. *Videnskabsselskabets Forhandlinger for 1863*, Christiania, 3 pp.

Klappa, C.F., Opalinski, P.R. & James, N.P. 1980: Middle Ordovician Table Head Group of Western Newfoundland, a revised stratigraphy. *Canadian Journal of Earth Sciences 17*, 1007–1019.

Kohut, J.J. 1972: Conodont biostratigraphy of the Lower Ordovician Orthoceras and Stein Limestones (3c), Norway. *Norsk Geologisk Tidsskrift 52*, 427–445.

Krill, A.G. & Zwann, B. 1987: Reinterpretation of Finnmarkian deformation on western *Norsk Geologisk Tidsskrift 67*, 15–24.

Lamont, A. & Lindström M. 1957: Arenigian and Llandeilian cherts identified in the Southern Uplands of Scotland by means of conodonts, etc. *Transactions of Edinburgh Geological Society 17*, 60–70.

Landing, E. 1976: Early Ordovician (Arenigian) conodont and graptolite biostratigraphy of the Taconic Allochthone, eastern New York. *Journal of Paleontology 50*, 614–646.

Lehnert, O. 1995a: Ordovizische Conodonten aus der Präkordillere Westargentiniens: Ihre Bedeutung für Stratigraphie und Paläeogeographie. *Erlanger Geologische Abhandlungen 125*, 1–193.

Lehnert, O. 1995b: Geodynamic processes in the Ordovician of the Argentine Precordillera: New biostratigraphic constraints. *In* Cooper, J.D., Droser, M.L. & Finney, S.C. (eds.): *Ordovician Odyssey: Short Papers for the Seventh International Symposium on the Ordovician System. SEPM, Pacific Section, Book 77*, 75–79.

Lindström, M. 1955a: Conodonts from the lowermost Ordovician strata of south-central Sweden. *Geologiska Föreningens i Stockholm Förhandlingar 76*, 517–604. Stockholm.

Lindström, M. 1955b: The conodonts described by A. R. Hadding, 1913. *Journal of Paleontology 29*, 105–111.

Lindström, M. 1960: A Lower-Middle Ordovician succession of conodont faunas. *21st International Geological Congress Reports 7*, 88–96. Copenhagen.

Lindström, M. 1964: Conodonts, 1–196. Elsevier, Amsterdam.

Lindström, M. 1970: A suprageneric taxonomy of the conodonts. *Lethaia 3*, 427–445.

Lindström, M. 1971: Lower Ordovician conodonts of Europe. *Geological Society of America Memoir 127*, 21–61. *In* Sweet W.C. and Bergström, S.M. (eds.): *Symposium on Conodont Biostratigraphy*.

Lindström, M. 1984: Baltoscandic conodont life environments in the Ordovician: sedimentologic and paleogeographic evidence. *In* Clark, D.L. (ed.): *Conodont Biofacies and Provincialism. Geological Society of America Special Paper 196*, 33–42.

Linnarsson, G. 1872: Anteckningar om den kambrisk-siluriska lagerserien i Jemtland. *Geologiska Föreningen i Stockholm Förhandlingar 1(3)*, 34–47.

Löfgren, A. 1978: Arenigian and Llanvirnian conodonts from Jämtland, northern Sweden. *Fossils and Strata 13*, 1–129.

Löfgren, A. 1985: Early Ordovician conodont biozonation at Finngrundet, south Bothnian Bay, Sweden. *Bulletin of the Geological Institution of the University of Uppsala 10*, 115–128.

Löfgren, A. 1990: Non-platform elements of the Ordovician conodont genus Polonodus. *Paläontologisches Zeitschrift 64*, 245–259.

Löfgren, A. 1993a: Conodonts from the lower Ordovician at Hunneberg, south-central Sweden. *Geological Magazine 130(2)*, 215–232.

Löfgren, A. 1993b: Arenig conodont succession from central Sweden. *Geologiska Föreningens i Stockholm Förhandlingar 115(3)*, 193–207.

Löfgren, A. 1994: Arenig (Lower Ordovician) conodonts and biozonation in the eastern Siljan district, central Sweden. *Journal of Paleontology 68(6)*, 1350–1368.

Löfgren, A. 1995a: The probable origin of the Ordovician conodont "*Cordylodus*" *horridus. Geobios 28*, 371–377.

Löfgren, A. 1995b: The middle Lanna/Volkhov Stage (middle Arenig) of Sweden and its conodont fauna. *Geological Magazine 132(6)*, 693–711.

Löfgren A. 1997a: Reinterpretation of the Lower Ordovician conodont apparatus *Paroistodus. Palaeontology 40(4)*, 913–929.

Löfgren, A. 1997b: Conodont faunas from the upper Tremadoc at Brattefors, south-central Sweden, and reconstruction of the *Paltodus* apparatus. GFF 119, 257–266.

Maletz, J., Löfgren, A. & Bergström, S. 1996: The base of the *Tetragraptus approximatus* Zone at Mt. Hunneberg, S.W. Sweden: a proposed global stratotype for the base of the second series of the Ordovician system. *Newsletter on Stratigraphy 34*, 129–159.

McCracken, A.D. 1989: *Protopanderodus* (Conodonta) from the Ordovician Road River Group, northern Yukon Territory, and the evolution of the genus. *Geological Survey of Canada, Bulletin 388*, 39 pp.

McCracken, A.D. 1991: Middle Ordovician conodonts from the Cordilleran Road River Group, northern Yukon Territory, Canada. *In* Orchard, M.J. & McCracken, A.D. (eds.): *Ordovician to Triassic Conodont Paleontology of the Canadian Cordillera. Geological Survey of Canada, Bulletin 417*, 41–54.

McHargue, T.R. 1982: Ontogeny, phylogeny and apparatus reconstruction of the conodont genus *Histiodella*, Joins Formation, Arbuckle Mountains, Oklahoma. *Journal of Paleontology 56*, 1410–1433.

McKerrow W.S., & Cocks, L.R.M. 1977: The location of the Iapetus Ocean suture in Newfoundland. *Canadian Journal of Earth Sciences 14*, 488–495.

McTavish, R.A. 1973: Prioniodontacean conodonts from the Emanuel Formation (Lower Ordovician) of Western Australia. *Geologica et Palaeontologica 7*, 27–58.

Meinich 1881: Dagbog fra en reise i Trysil sommeren 1879 samt om Kvitvola-etagens forhold til Trysilfjeldets kvartsit og sandstene. *Nyt Magazin for Naturvidenskaberne 26*, 12–33.

Merrill, G.K. 1980: Ordovician conodonts from the Åland Islands, Finland. *Geologiska Föreningens i Stockholm Förhandlingar 101*, 329–341.

Miller, J.F. 1980: Taxonomic revisions of some Upper Cambrian and Lower Ordovician conodonts, with comments on their evolution. *University of Kansas Paleontological Contributions Paper 99*, 44 pp.

Müller, K.J. 1962: Taxonomy, evolution and ecology of conodonts. *In* Hass, W.H., *et al.* (eds.) *Treatise on Invertebrate Paleontology. - Part W. Miscellanea*, W83-W91. Geological Society of America and University of Kansas Press, Boulder.

Müller, K.J. & Hinz, I. 1991: Upper Cambrian conodonts from Sweden. *Fossils and Strata 28.* 153 pp.

Münster, T. 1891: Foreløbige meddelelser fra reiser i Mjøsegnene, udførte for den geologiske undersøgelse sommeren 1889. *Norges Geologiske Undersøgelse 1, Aarbog for 1891*, 11–18.

Münster T. 1900: Kartbladet Lillehammer. *Norges Geologiske Undersøgelse 30*, 49 pp.

Männik, P. & Viira, V. 1990: Conodonts. *In* Kaljo, D. (ed.): *IUGS, IGCP Field Meeting, Estonia 1990. An Excursion Guidebook*, 148–153. Institute of Geology, Estonian Academy of Sciences.

Neumann, R.B. 1984: Geology and paleobiology of islands in the Ordovician Iapetus Ocean: Review and implications. *Geological Society of America, Bulletin 95*, 1188–1201.

Neumann, R.B. & Bruton, D.L. 1974: Early Middle Ordovician fossils from the Hølonda area, Trondheim Region, Norway. *Norsk Geologisk Tidsskrift 54*, 69–115.

Nickelsen, R.P., Hossack, J.R., Garton, M. & Repetski, J. 1985: Late Precambrian to Ordovician stratigraphy and correlation in the Valdres and Synnfjell thrust sheets of the Valdres area, southern Norwegian Caledonides, with some comments on sedimentation. *In* Gee, D.G. & Sturt, B.A. (eds.): *The Caledonide Orogen - Scandinavia and Related Areas*, 369–378. John Wiley & Sons, Chichester.

Nicoll, R.S. 1992: Evolution of the conodont genus *Cordylodus* and the Cambrian-Ordovician boundary. *In* Webby, B.D. & Laurie, J. (eds.): *Global Perspectives on Ordovician Geology*, 105–113. A.A. Balkema, Brookfield.

Nielsen, A.T. 1995: Trilobite systematics, biostratigraphy and palaeoecology of the Lower Ordovician Komstad Limestone and Huk Formations, southern Scandinavia. *Fossils and Strata 38*, pp. 1–374.

Nowlan, G.S. 1981: Some Ordovician conodont faunules from the Miramichi Anticlinorium, New Brunswick. *Geological Survey of Canada, Bulletin 345*, 35 pp.

Nowlan, G.S. 1983: Biostratigraphic, paleogeographic, and tectonic implications of Late Ordovician conodonts from the Grog Brook Group, northwestern New Brunswick. *Canadian Journal of Earth Sciences 20*, 651–671.

Nystuen, J.P. 1974: ENGERDAL, berggrunnsgeologisk kart 2018 I-M. 1:50000.

Nystuen, J.P. 1975: ELVDAL, berggrunnsgeologisk kart 2018 III-M. 1:50000.

Nystuen, J.P. 1981: The late Precambrian "Sparagmites" of southern Norway, a major Caledonian Allochthon - the Osen-Røa Nappe Complex. *American Journal of Science 281*, 69–94.

Olgun, O. 1987: Komponenten-Analyse und Conodonten-Stratigraphie der Orthoceratenkalksteine im Gebiet Falbygden, Västergötland, Mittelschweden. *Sveriges Geologiska Undersökning Ca 70*, 78 pp.

Orchard, M.J. 1980: Upper Ordovician conodonts from England and Wales. *Geologica et Paleontologica 14*, 9–44.

Owen, A.W., Bruton, D.L., Bockelie, J.F. & Bockelie, T.G. 1990: The Ordovician successions of the Oslo Region, Norway. *Norges Geologiske Undersøkelse, Special Publication 4*, 3–54.

Palm, H., Gee, D.G., Dyrelius, D. & Bjørklund, L. 1991: A reflection seismic image of Caledonian structure in Central Sweden. *Sveriges Geologiska Undersökning Ca 75*, 36 pp.

Pander, H.C. 1856: Monographie der fossilen Fische des Silurischen Systems der Russisch-Baltischen Gouvernements. *Buchdruckerei der Kaiserlichen Akademie der Wissenschaften*, St. Petersburg, 1–91. St. Petersburg.

Pohler, S. & Orchard, M.J. 1990: Ordovician conodont biostratigraphy, Western Canadian Cordillera. *Geological Survey of Canada, Paper 90–15*, 37 pp.

Pohler, S.L., Barnes, C.R. & James, N.P. 1987: Reconstructing a lost faunal realm: conodonts from mega-conglomerates of the Ordo-vician Cow Head Group, western Newfoundland. *In* Austin, R.L. (ed.): *Conodonts: Investigative Techniques and Applications. British Micro-palaeontological Society Series*, 341–362. Ellis Horwood Limited, Chichester.

Põldvere, A., Meidla, T., Bauert, G., Bauert, H. & Stouge, S. 1998: Tartu (453) Drillcore. *Estonian Geological Sections Bulletin 1.* 48 pp.

Rasmussen, J.A. 1991: Conodont stratigraphy of the Lower Ordovician Huk Formation at Slemmestad, southern Norway. *Norsk Geologisk Tidsskrift 71,* 265 – 288.

[Rasmussen, J.A. 1994: Stratigraphy, conodont faunas and depositional setting of the Lower Allochthonous limestones within the Scandinavian Caledonides. Unpublished PhD thesis, University of Copenhagen 1994. 335 pp.]

Rasmussen, J.A. 1998: A reinterpretation of the conodont Atlantic Realm in the late Early Ordovician (early Llanvirn). *In* Szaniawski, H. (ed.): Proceedings of the Sixth European Conodont Symposium (ECOS VI). *Palaeontologica Polonica 58,* 67 – 77.

Rasmussen, J.A. & Bruton, D.L. 1994: Stratigraphy of Ordovician limestones, Lower Allochthon, Scandinavian Caledonides. *Norsk Geologisk Tidsskrift 74,* 199 – 212.

Rasmussen, J.A. & Stouge, S.S. 1988: Conodonts from the Lower Ordovician Stein Limestone, Norway, and a correlation across Iapetus. *In* Williams, S.H. & Barnes, C.R. (eds.): *Fifth International Symposium on the Ordovician System, Program and Abstracts,* 78.

Rasmussen, J.A. & Stouge, S.S. 1989: Middle Ordovician conodonts from allochthonous limestones at Høyberget, southeastern Norwegian Caledonides. *Norsk Geologisk Tidsskrift 69,* 103 – 110.

Rasmussen, J.A. & Stouge, S. 1995: Late Arenig-Early Llanvirn conodont biofacies across the Iapetus Ocean. *In* Cooper, J.D., Droser, M.L. & Finney, S.C. (eds.): Ordovician Odyssey: Short Papers for the Seventh International Symposium on the Ordovician System. *SEPM, Pacific Section, Book 77,* 443 – 447.

Rhodes, F.H.T. 1955: The conodont fauna of the Keisley Limestone. *Geological Society of London, Quarterly Journal 111,* 117 – 142.

Roberts, D. & Gee, D.G. 1985: An introduction to the structure of the Scandinavian Caledonides. *In* Gee, D.G. & Sturt, B.A. (eds.): *The Caledonide Orogen-Scandinavia and Related Areas.* John Wiley & Sons, Chichester.

Robison, R.A. (ed.) 1981: *Treatise on Invertebrate Paleontology, pt. W, Miscellanea, suppl. 2, Conodonta. Geological Society of America and University of Kansas Press, Boulder.* 202 pp.

Ross, R.J. (Jr), Ethington, R.L. & Mitchell, C.E. 1991: Stratotype of Ordovician Whiterock Series. *Palaios 6,* 156 – 173.

Röshoff, K. 1978: Structures of the Tännäs augen gneiss Nappe and its relation to upper- and overlying units in the central Scandinavian Caledonides. *Sveriges Geologiske Undersökning, Serie C, 739,* 35 pp.

Sansom, I.J., Armstrong, H.A. & Smith M.P. 1994: The apparatus architecture of *Panderodus* and its implications for coniform conodont classification. *Palaeontology 37(4),* 781 – 799.

Sansom, I.J., Smith, M.P., Armstrong, H.A. & Smith M.M. 1992: Presence of the earliest vertebrate hard tissues in conodonts. *Science 256,* 1308 – 1311.

Schiøtz, O.E. 1874: Beretning om nogle undersøkelser over Sparagmit-Kvarts-Fjeldet i den Østlige del af Hamar Stift. *Nyt Magazin for Naturvidenskaberne 20,* 25 – 123.

Schiøtz, O.E. 1883: Sparagmit-Kvarts-Fjeldet i den østlige del af Hamar Stift. *Nyt Magazin for Naturvidenskaberne 27,* 154 – 216.

Schiøtz, O.E. 1892: Sparagmit-Kvarts-Fjeldet langs Grænsen i Hamar Stift og i Herjedalen. *Nyt Magazin for Naturvidenskaberne 32,* 1 – 98.

Scotese, C.R. & McKerrow, W.S. 1990: Revised world maps and introduction to memoir. *In* McKerrow, W.S. & Scotese, C.R. (eds.): *Palaeozoic Palaeogeography and Biogeography. Geological Society, London, Memoir 12,* 1 – 21.

Sergeeva, S.P. 1963: [Conodonts from the Lower Ordovician in the Leningrad region.] *Paleontologicheshij Zhurnal 1963* 93–108 (Akademia Nauk SSSR) (In Russian).

Sergeeva, S.P. 1964: [On the stratigraphic significance of the Lower Ordovician conodonts of the Leningrad region.] *Leningrad Univ. Vestnik, Geologiya i Geografii 12,* 56 – 60. (In Russian).

Sergeeva, S.P. 1974: [Some new conodonts from Ordovician strata in the Leningrad region.] *Paleontologicheshiy Sbornik 11(2),* 79 – 84. (In Russian).

Serpagli, E. 1967: I conodonti dell'Ordoviciano superiore (Ashgilliano) delle Alpi Carniche. *Bolletino della Società Paleontologica Italiana 6,* 30 – 111.

Serpagli, E. 1974: Lower Ordovician conodonts from Precordilleran Argentina (Province of San Juan). *Bolletino della Società Paleontologica Italiana 13,* 17 – 98.

Skjeseth, S. 1952: On the Lower Didymograptus Zone (3b) of Ringsaker and contemporaneous deposits in Scandinavia. *Norsk Geologisk Tidsskrift 30,* 138 – 182.

Skjeseth, S. 1963: Contributions to the geology of the Mjøsa Districts and the classical sparagmite area in southern Norway. *Norges Geologiske Undersøkelse 220,* 1 – 126.

Smith, M.P. 1982: Conodonts from the Ordovcian of East Greenland. *Rapport Grønlands geologiske Undersøgelse 108,* 14.

Smith, M.P. 1991: Early Ordovician conodonts of East and North Greenland. *Meddelelser om Grønland, Geoscience 26,* 81 pp.

Spjeldnæs, N. 1985: Biostratigraphy of the Scandinavian Caledonides. *In* Gee, D.G. & Sturt, B.A. (eds.): *The Caledonide Orogen - Scandinavia and Related Areas,* 317 – 319. John Wiley & Sons Limited, Chichester.

Stait, K., & Druce, E.C. 1993: Conodonts from the Lower Ordovician Coolibah Formation, Georgina Basin, central Australia. *Journal of Australian Geology and Geophysics 13,* 293 – 322.

Stauffer, C.R. 1935: Conodonts of the Glenwood beds. *Geological Society of America Bulletin 46,* 125 – 168.

Stenzel, S.R., Knight, I. & James, N.P. 1990: Carbonate platform to foreland basin: revised stratigraphy of the Table Head Group (Middle Ordovician), western Newfoundland. *Canadian Journal of Earth Science 27,* 14 – 26.

Stephens, M.B. 1988: The Scandinavian Caledonides: a complexity of collisions. *Geology Today, January-February,* 20 – 26.

Stephens, M.B., Kullerud, K. & Claesson, S. 1993: Early Caledonian tectonothermal evolution in ourboard terranes, central Scandinavian Caledonides: new constraints from U-Pb zircon dates. *Journal of the Geological Society 150,* 51 – 56.

Stouge, S.S. 1975: Conodontzonerne i orthoceratitkalken (Nedre Ordovicium) på Bornholm og i Fågelsång. *Dansk geologisk Forening Årsskrift 1974,* 32 – 38.

Stouge, S.S. 1980a: Conodonts from the Davidsville Group, northeastern Newfoundland. *Canadian Journal of Earth Sciences 17,* 268 – 272.

Stouge, S.S 1980b: Lower and Middle Ordovician conodonts from Central correlatives in western Newfoundland. *Mineral Development Division, Department of Mines and Energy, Government of Newfoundland and Labrador, Report 80 – 1,* 16 pp.

Stouge, S.S. 1982: Preliminary Conodont Biostratigraphy and Correlation of Lower to Middle Ordovician Carbonates of the St. George Group, Great Northern Peninsula, Newfoundland. *Mineral Development Division, Department of Mines and Energy, Government of Newfoundland and Labrador, Report 82 – 3,* 59 pp.

Stouge, S.S. 1984: Conodonts of the Middle Ordovician Table Head Formation, western Newfoundland. *Fossils and Strata 16,* 145 pp.

Stouge, S.S. 1989: Lower Ordovician (Ontikan) conodont biostratigraphy in Scandinavia. *Proceedings of the Academy of Sciences of the Estonia SSR Geologia 38(2)*.

Stouge, S.S. & Boyce, W.D. 1983: Fossils of Northwestern Newfoundland and Southwestern Labrador: conodonts and trilobites. *Newfoundland Department of Mines & Energy, Mineral Dev. Division, Report 83–3*, 55 pp.

Stouge, S. & Bagnoli, G. 1988: Early Ordovician Conodonts from Cow Head Peninsula, Western Newfoundland. *Palaeontographia Italica 75*, 89–179.

Stouge, S. & Bagnoli, G. 1990: Lower Ordovician (Volkhovian-Kundan) conodonts from Hagudden, northern Öland, Sweden. *Palaeontographica Italica 77*, 1–54.

Stouge S. & Bagnoli, G. 1997: The Ledge Section, Cow Head Peninsula, western Newfoundland: a candidate as suitable GSSP section: Discussion. *Ordovician News 14*, 38–43.

Strand, T. 1929: The Cambrian beds of the Mjøsa district in Norway. *Norsk Geologisk Tidsskrift 10*, 308–365.

Strand, T. 1938: Nordre Etnedal. Beskrivelse til det geologiske gradteigskart. *Norges Geologiske Undersøkelse 152*, 71 pp.

Strand, T. 1954: Aurdal. *Norges Geologiske Undersøkelse 187*. 71 pp.

Sturkell, E. 1991: Tremadocian Ceratopyge Limestone identified by means of conodonts, in Jämtland, Sweden. *Geologiska Föreningens i Stockholm Förhandlingar 113*, 185–188.

Størmer, L. 1953: The middle Ordovician of the Oslo Region, Norway. No. 1. stratigraphy. *Norsk Geologisk Tidsskrift 31*, 37–141.

Størmer, L. 1967: Some aspects of the Caledonian geosyncline and foreland west of the Baltic Shield. *Quaternary Journal of the Geological Society of London 123*, 183–214.

Sweet, W.C. 1979: Late Ordovician conodonts and biostratigraphy of the western midcontinent Province. *Brigham Young University Geological Studies 26(3)*, 45–86.

Sweet, W.C. 1984: Graphic correlation of upper Middle and Upper Ordovician rocks, North American Midcontinent Province, U.S.A. *In* Bruton, D.L. (ed.): *Aspects of the Ordovician System*, 23–35. *Palaeontological Contributions from the University of Oslo 295*.

Sweet, W.C. 1988: *The Conodonta: Morphology, Taxonomy, Paleoecology and Evolutionary History of a Long-Extinct Animal Phylum.* Oxford Monographs on Geology and Geophysics *10*, 212 pp.

Sweet, W.C. & Bergström, S.M. 1962: Conodonts from the Pratt Ferry Formation (Middle Ordovician) of Alabama. *Journal of Paleontology 36*, 1214–1252.

Sweet, W.C. & Bergström, S.M. 1972: Multielement taxonomy and Ordovician conodonts. *Geologica et Palaeontologica, Special Volume 1*, 29–42.

Sweet, W.C. & Bergström, S.M. 1984: Conodont provinces and biofacies of the Late Ordovician. *In* Clark, D.L. (ed.): *Conodont Biofacies and Provincialism. The Geological Society of America, Special Paper 196*, 69–87.

Sweet, W.C. & Ethington R.L. & Barnes, C.R. 1971: North America Middle and Upper Ordovician conodont faunas. *Geological Society of America Memoir 127*, 163–193.

Thorslund, P. 1937: Kvartsiter, sandstenar och tektonik inom Sunneområdet i Jämtland. *Sveriges Geologiska Undersökning C 409*, 30 pp.

Tipnis, R.S., Chatterton, B.D.E. & Ludvigsen, R. 1979: Biostratigraphy of Ordovician conodonts from the southern Mckenzie Mountains. *In* Stelck, C.R. & Chatterton, B.D.E. (eds.): *Western and Arctic Canadian biostratigraphy. Geological Association of Canada, Special Paper 18*, 39–91.

Tjernvik, T.E. & Johansson, J.V. 1980: Description of the upper portion of drill-core from Finngrundet in the South Bothnian Bay. *Bulletin of the Geological Institution of the University of Uppsala, N.S. 8*, 173–204.

Torsvik, T.H., Ryan, P.D., Trench, A. & Harper, D.A.T. 1991: Cambrian-Ordovician paleogeography of Baltica. *Geology 19*, 7–10.

Trench, A. & Torsvik, T.H. 1992: Palaeomagnetic constraints on the Early-Middle Ordovician palaeogeography of Europe: Recent advances. *In* Webby, B.D. & Laurie, J.R. (eds.): *Global Perspectives on Ordovician Geology*, 255–259. A.A. Balkema, Brookfield.

Törnebohm, A.E. 1896: Grunddragen af det centrala Scandinaviens bergbyggnad. *Kungliga Svenska Veteskap-Akademins Handlingar 28*, 212 pp.

Urbanek, Z. & Baranowski, Z. 1986: Revision of age of the Radzimowice slates from Góry Kaczawskie Mts. (western Sudetes, Poland) based on conodonts. *Annales de la Societe Geologique de Pologne 56*, 399–408.

Uyeno, T.T. & Barnes, C.R. 1970: Conodonts from the Lévis Formation (Zone D1) (Middle Ordovician), Lévis, Quebec. *Geological Survey of Canada, Bulletin 187*, 99–123.

Van Wamel, W.A. 1974: Conodont biostratigraphy of the Upper Cambrian and Lower Ordovician of north-western Öland, south-eastern Sweden. *Utrecht Micropaleontological Bulletins 10*, 1–126.

Viira, V. 1966: [Distribution of conodonts in the Lower Ordovician sequence of Suhkrumägi (Tallinn).] *Eesti NSV Teaduste Akadeemia Toimetised, Keemia, Geologia 15*, 150–155 (in Russian).

Viira, V. 1967: [Ordovician conodont succession in the Ohesaare core.] *Eesti NSV Teaduste Akadeemia Toimetised, Keemia, Geologia 16*, 319–329 (in Russian).

Viira, V. 1972: On symmetry of some Middle Ordovician conodonts. *Geologica et Paleontologica 6*, 45–49.

Viira, V. 1974: [Ordovician conodonts of the east Baltic.] *Geological Institute of the Academy of Sciences of the Estonian S.S.R., "Valgus", Tallinn.* 142 pp. (In Russian).

Waern, B., Thorslund, P. & Henningsmoen, G. 1948: Deep boring through Ordovician and Silurian strata at Kinnekulle, *Västergötland. Bulletin of the Geological Institution of the University of Uppsala 32*, 374–432.

Wandås, B.T.G. 1983: The Middle Ordovician of the Oslo Region, Norway, 33. Trilobites from the lowermost part of the Ogygiocaris Series. *Norsk Geologisk Tidsskrift 63*, 211–267.

Webby, B.D. 1984: Ordovician reefs and climate: a review. *In* Bruton, D.L. (ed.): *Aspects of the Ordovician System*, 89–100. *Palaeontological Contributions from the University of Oslo 295*.

Webers, G.F. 1966: The Middle and Upper Ordovician Conodont Faunas of Minnesota. *Minnesota Geological Survey, Special Publication 4*, 123 pp.

Williams, S.H. 1984: Lower Ordovician graptolites from Gausdal, central southern Norway: a reassessment of the fauna. *Norges Geologiske Undersøkelse 395*, 1–23.

Williams, S.H. & Stevens, R.K. 1988: Early Ordovician (Arenig) graptolites of the Cow Head Group, western Newfoundland, Canada. *Palaeontographica Canadiana 5*. 167 pp.

Williams, S.H., Barnes, C.R., O'Brien, F.H.C. & Boyce, W.D. 1994: A proposed global stratotype for the second series of the Ordovocian System: Cow Head Peninsula, western Newfoundland. *Bulletin of Canadian Petroleum Geology 42*, 219–231.

Wiman, C. 1893: Ueber die Silurformation in Jemtland. *Bulletin of the Geological Institutions of the University of Uppsala 1*, 256–276.

Wiman, C. 1898: Kambrisch-silurische Faciesbildungen in Jemtland. *Bulletin of the Geological Institution of the University of Upsala III*, 269–304.

Wolska, Z. 1961: Konodonty z ordowickich glazow narzutowych Polski [Conodonts from Ordovician erratic boulders of Poland]. *Acta Palaeontologica Polonica 6*, 339–365.

Worsley, D., Aarhus, N., Basset, M.G., Hove, M.P.A., Mørk, A. & Olaussen, S. 1983: The Silurian succession of the Oslo Region. *384*, 1–57.

Zhang, J. 1997: The Lower Ordovician conodont *Eoplacognathus crassus* Chen & Zhang, 1993. GFF *119*, 61–65.

Zhang, J. 1998a: The Ordovician conodont genus *Pygodus. In* Szaniawski, H. (ed.): Proceedings of the Sixth European Conodont Symposium (ECOS VI). *Palaeontologica Polonica 58*, 87–105.

Zhang, J. 1998b: Conodonts from the Guniutan Formation (Llanvirnian) in Hubei and Hunan Provinces, south-central China. *Stockholm Contributions in Geology 46*, 1–161.

Zhang, J. & Sturkell E.F.F. 1998: Aserian and Lasnamägian (Middle Ordovician) conodont biostratigraphy and lithology at Kullstaberg and Lunne in Jämtland, central Sweden. GFF *120*, 75–83.

Ziegler, W. (ed.) 1977: *Catalogue of Conodonts III*, 1–574. Stuttgart.

Appendix 1

The content of calcium-carbonate, magnesium, strontium and manganese within the studied sections. The geochemical parameters were determined with an atomic absorption spectrophotometer.

Locality	Sample	CaCO3%	Ca%	Sr ppm	Mg ppm	Mn ppm
Steinsodden	69601	70.10	39.40	640	4080	2100
	69610	74.50	39.60	341	2940	1480
	69614	75.30	39.50	345	3730	1310
	69619	70.20	39.50	331	3870	1560
	69622	82.10	39.60	324	3290	1830
	69626	65.60	39.40	359	4630	1220
	69628	70.10	39.50	298	3530	1210
	69629	80.80	39.60	443	3190	2080
	69631	90.00	39.60	317	2770	1450
	69633	83.90	39.60	398	2900	1830
	69635	80.60	39.60	323	3160	1120
	69638	88.50	39.60	389	2970	1940
	69642	83.80	39.60	284	3300	1140
	69645	68.90	39.40	290	4240	1190
	69648	78.70	39.50	326	3650	1220
	69649	84.60	39.50	323	3740	1060
	69653	80.50	39.50	363	3850	2120
	69656	36.10	38.70	382	9830	1100
	69659	59.60	39.20	415	6170	1940
	69662	58.70	39.10	410	6660	1760
	69667	54.70	38.80	467	8830	1800
	69672	71.80	39.30	557	4800	2080
	69671	52.80	38.80	436	8420	1600
	69673	47.10	38.50	513	11100	1350
	69675	76.00	39.10	641	6380	1760
	69676	43.60	38.50	961	10300	1490
Andersön-A	97802	16.48	6.60	70	1800	490
	97811	60.93	24.40	160	1900	1150
	97814	77.91	31.20	240	1900	1160
	97815	75.16	30.10	210	1900	1280
	97816	80.15	32.10	240	1900	1330
	97818	83.15	33.30	270	2000	1440
	97821	85.65	34.30	310	2200	1300
	97823	81.40	32.60	220	2100	1200
	97825	80.15	32.10	280	2100	1580
	97826	83.15	33.30	260	2000	1370
	97829	69.67	27.90	180	1800	1870
	97830	85.40	34.20	240	1800	2220
	97832	89.89	36.00	380	1900	2240
	97833	94.64	37.90	370	2100	2140
	97834	86.40	34.60	250	2500	1970
	97835	75.16	30.10	240	2400	2380
	97836	89.89	36.00	230	2400	2230
	97837	79.90	32.00	240	2400	2180
Andersön-B	99451	75.66	31.10	220	2400	1440
	99453	76.91	30.80	230	3100	1370
	99455	83.40	33.40	210	2100	1260
	99457	84.40	33.80	160	1900	2280
	99459	82.15	32.90	250	2200	1560
	99461	75.66	30.30	230	2400	2210
	99462	86.65	34.70	380	2500	1930
	99463	89.64	35.90	530	2400	1950
	99465	90.14	36.10	280	2100	2450
	99466	79.65	31.90	220	2100	2660
	99468	80.65	32.30	210	2500	2480
	99470	86.15	34.50	300	2100	2120
	99472	84.65	33.90	310	2500	2110

Locality	Sample	CaCO3%	Ca%	Sr ppm	Mg ppm	Mn ppm
	99474	79.40	31.80	240	2700	2590
	99475	57.43	23.00	210	3100	2360
	99477	79.90	32.00	340	2700	3250
	99478	81.90	32.80	330	2700	3130
	99480	72.66	29.10	330	3500	1850
	99481	76.91	30.80	460	3500	5480
	99483	76.91	30.80	440	3200	7000
	99485	74.66	29.90	620	3300	9200
Engerdal	97652	84.15	33.70	270	2000	4230
	97655	82.15	32.90	310	2200	6000
	97656	89.39	35.80	260	2000	4310
Hestekinn	97711	60.68	24.30	260	2500	1230
	97713	0.70	0.28	0	350	0
	97714	49.19	19.70	310	2000	1450
	97715	69.42	27.80	330	2000	1460
	97717	44.95	18.00	270	2300	1210
Jøronlia	97718	71.60	28.50	800	3800	1980
	97720	65.42	26.20	590	3000	1920
	97723	16.98	6.80	90	2900	670
	97727	4.05	1.62	0	500	100
	97729	71.41	28.60	580	2500	2100
	97731	43.20	17.30	430	3200	1280
	97732	1.77	0.71	0	4700	260
	97734	39.20	15.70	400	3700	1180
Røste	97759	69.92	28.00	500	2300	2230
	97762	63.17	25.30	260	3500	11740
	97764	45.20	18.10	210	2800	1230
	97766	71.41	28.60	450	2300	2030
	97768	81.90	32.80	890	2600	2890
Herram	97697	43.95	17.60	250	3100	1550
Høyberget	99510	50.19	20.10	150	3600	5300
Sorken	97791	79.15	31.70	190	1700	4360
Grøslii	97664	81.40	32.60	1370	2400	840

Appendix 2

Distribution of conodont elements in the Anderson-A section

Anderson - A

Species/Elements	Sample 97xxx	799	800	802	804	805	807	810	812	815	817	819	822	825	827	829	831	832	833	834	837	Sum
Acodus deltatus	P	1																				1
	Sb					1																1
Lundodus gladiatus	Sb					4																4
Ansella jemtlandica	M																				29	29
	Sa																				11	11
	Sb																				7	7
	Sc																				13	13
Baltoniodus medius	Pa																			2	2	4
	Pb																			4	6	10
	M																			10	1	11
	Sa																			3		3
	Sb																			5	4	9
	Sc																			4	3	7
	Sd																				7	7
Baltoniodus navis	Pa												1									1
	Pb								1		2	2	5									10
	M								1				6	5								12
	Sa						1				1		2									4
	Sb												3									3
	Sc										2		6									8
	Sd										2		2									4
Baltoniodus norrlandicus	Pa														10	6	1	11	1			29
	Pb														25	6	15	26	2			74
	M														17	4	13	22	4			60
	Sa														6		4	7	2			19
	Sb														8	5	2	15	2			32
	Sc														14	4	10	13	4			45
	Sd														10	5	11	24	1			51
Coelocerodontus? sp.																					2	2
Cornuodus longibasis	Drep.					1			4				7	4				3			10	34
	oist.											1										4
	costate drep.																					1
Costiconus costatus											4	1	3	3			2					13
Dapsilodus? viruensis	acont.																			3	4	7
	acod.																			1	1	1

Andersön - A

Species/Elements	Sample 97xxx	799	800	802	804	805	807	810	812	815	817	819	822	825	827	829	831	832	833	834	837	Sum
Diaphorodus tovei	Sa		1																			1
	Sb		1																			1
Drepanodus arcuatus	arcuat.	2						2	2								5					11
	gracili.																1	1				2
	oist.						1								1		1					4
	pipa.	1							1				1									4
	sculp.	1						3						2	3		4	3			9	25
Drepanodus planus	arcuat.		3			1					1									1		6
	pipa.		1		2																	3
	sculp.		1																			1
Drepanoistodus basiovalis	suberect.												4		2							6
	drep. 1												1		1							2
	drep. 2												5		5							10
	drep. 3												17		15							32
	oist.												11		8			2				21
Drepanoistodus forceps	suberect.	1			2			8	2		1											14
	drep. 1				1	2		3	2		2											10
	drep. 2	1		1		2		12	3	1	5											25
	drep. 3			3	5	10	3	52	20	4	31											128
	oist.	1	3	3	5	18	2	42	30		44											148
Drepanoistodus stougei	suberect.															1	4	1				6
	drep. 1															2	7	2				11
	drep. 2															3	3	2				8
	drep. 3															22	12	15	2			51
	oist.															10	14	6	1			31
"Drepanoistodus venustus"	drep. 2																			1		1
	drep. 3																			3		3
	oist.																			5		5
Drepanoistodus sp. indet.	drep. 2											1										1
	drep. 3											1		4								5
Histiodella holodentata	P																				3	3
Lenodus cf. variabilis	Pb																		4			4
	Sa																				1	1
	Sd																				3	3
Lenodus sp. A	Pb												3									3
	Sb												2									2
Lenodus sp. B	Sb																	2				2
	Sd																3					3
Lenodus sp. indet.	Pa															1						1

Taxon	Element									Total
Microzarkodina flabellum	P						2		1	5
	M						2	2		2
	Sa							1		1
	Sb					1		1		2
	Sc					2		1		3
	Sd					2				3
Microzarkodina hagetiana	P	4						1		4
	M	1								1
	Sa	2								2
	Sb	2								2
	Sc	4								4
Microzarkodina ozarkodella	P	17	1							18
	M	15								15
	Sa	1								1
	Sb	2								2
	Sc	5								5
	Sd	1								1
Microzarkodina parva	P	17	16	8	4	5	1			51
	M	8	7	3	7	1	2			28
	Sa	2	1	2	4					9
	Sb	1			1					3
	Sc	6	1	3	2	3				14
	Sd	3			1	1				7
Oelandodus elongatus	P			8				1		3
	M			3				1		3
	S			2				5	21	33
Oepikodus evae	Pa									5
	Pb									9
	M									4
	Sb									3
	Sd									2
Oistodus lanceolatus	Sa			1				1		1
	Sb			1		3		1		5
	Sc					1				1
Paltodus subaequalis	drep.							1		1
Paracordylodus gracilis	P			24				6		6
	M			53				27	22	49
	S							104	48	152
Paroistodus horridus	drep.	2								2
Paroistodus originalis	drep.		4	24	10	23	29	21		118
	oist.		3	53	13	17	20	24		135

Andersön - A

Species/Elements	Sample 97xxx	799	800	802	804	805	807	810	812	815	817	819	822	825	827	829	831	832	833	834	837	Sum
Paroistodus parallelus	drep.	13	10	2		3																28
	oist.			2		1																3
Paroistodus proteus	drep.	27	18		3																	48
	oist.	6	9		1																	16
Periodon flabellum	Pa								11													11
	Pb								16													16
	M								22													22
	Sa								6													6
	Sb								5													5
	Sc							1	23													24
	Sd								7													7
Periodon macrodentata	Pa																		1		11	12
	Pb																			2	3	5
	M																		1	1	12	14
	Sa																		1	1	3	5
	Sb																			1	7	8
	Sc																		1		11	12
	Sd																				7	7
Periodon selenopsis	Pa					2																2
	Pb					2																2
	M					5																5
	Sb					1																1
	Sc					4																4
Periodon zgierzensis	Pa								4		2		17			3						26
	Pb								1				5			1						7
	M								1		3		26			7						37
	Sa												5									5
	Sb										1		6			1						8
	Sc										2	1	12			3						18
	Sd										2		16									18
Polonodus? cf. *tablepointensis*	Pa-1																				2	2
	Pa-2																				1	1
	Pb																				6	6
	M																				1	1
	Sb																				5	5
	Sd																				3	3
Polonodus sp. indet.	Pa-2																		1	1		2
	Pb																			2		2
	Sb																		1			1
	Sd																			1		1

Taxon	Element	234	153	16	26	104	12	177	267	14	181	25	322	116	247	129	248	315	61	115	311	Total
Prioniodus elegans	Pa																					2
	Pb																					2
	M																					3
	Sb																					1
Protopanderodus calceatus - P. sulcatus complex	scand.	1																1				4
	acont.	4							2									3			2	27
Protopanderodus rectus	scand.	1					3	4		1	10		11	6	6	7	7					57
	sym. acont.	1	1				13	29			22		30	8	13	10	19	19				151
	asym. acont.	1	1			3	6	17			14		15	7	6	5	19	19				99
Protopanderodus robustus	scand.																				5	5
	sym. acont.																				39	39
	asym. acont.																				9	9
Protoprioniodus cf. costatus	P	1																				1
	Sb	2																				2
Scalpellodus gracilis	scand.																	4	4		2	10
	longb. drep.																3	8	20		6	37
	shortb. drep.																	4	8		11	23
Scalpellodus latus	scand.								1				1	2	5	3	6	12		1		30
	longb. drep.													4	24	11	20	80		1		140
	shortb. drep.												2	5	7	5	15	31				65
Scolopodus quadratus								1														1
"Scolopodus" peselephantis														1	1	1	2				1	5
Semiacontiodus cornuformis	cornuf. A																	1				1
	cornuf. B																			2		2
	drep.													1		2		10		10		23
Stolodus stola	P				1																	1
	Sb				2																	2
	Sc		1		6																	7
	Sd		1		4																	5
Tetraprioniodus robustus	Pa																					13
	Pb																					5
	M																					1
	Sa																					3
	Sb																					1
	Sd																					6
Triangulodus brevibasis	P										3			4								7
	M													5								5
	Sa										1											1
	Sb										1			2								3
	Sd													2								2
Unidentifiable		6						2						1				1	1		1	18
Total		234	153	16	26	104	12	177	267	14	181	25	322	116	247	129	248	315	61	115	311	3073

Appendix 3

Distribution of conodont elements in the Andersön-B and Andersön-C sections.

Andersön - B Species/Elements	Sample 99xxx	451	452	453	454	455	456	457	458	459	460	461	462	463	464	465	466	467	468
Amorphognathus? sp.	Pa																		
Ansella jemtlandica	P																	32	
	M																	15	
	Sa																	10	
	Sb																	9	
	Sc														1			7	
Baltoniodus medius	Pa														1	3	2	2	1
	Pb														6	15	2	5	2
	M														4	4	2	3	6
	Sa														2	7			1
	Sb														3	4	1	2	1
	Sc														2	8		4	4
	Sd														2	4	1	3	3
Baltoniodus navis	Pa						1	1	4	3	1								
	Pb							2	11	6	1								
	M						1	1	14		2								
	Sa								3	1									
	Sb							1	9	1	3								
	Sc								6	2	1								
	Sd						1	1	5	3	3								
Baltoniodus norrlandicus	Pa										4	18	4						
	Pb										6	33	11						
	M										12	16	5						
	Sa										6	13	3						
	Sb										7	17	4						
	Sc										7	26	5						
	Sd										10	26	2						
Baltoniodus prevariabilis	Pa																		
	Pb																		
	M																		
	Sa																		
	Sb																		
	Sc																		
	Sd																		
Baltoniodus sp. indet.	Pa																		
Baltoniodus? sp.	M					1													
Cornuodus longibasis						1		3	3		3	11			1	3	1	4	4
Costiconus costatus	drep.						1		1		2	9			6				1
	oist.										2	6							
	costate						1		2		4	5				3			1
Costiconus ethingtoni	drep.																		
	oist.																		
	costate																		
Costiconus cf. *ethingtoni*	drep.																		
	oist.																		
	costate																		
Costiconus sp. A	drep.																		
	oist.																		
	costate																		
Costiconus sp. B	oist.																		
Costiconus sp. indet.	drep.																		
	costate																		

469	470	471	472	473	474	475	476	477	478	479	480	481	484	485	486	488	491	Total	Andersön - C	492	496	Total
														1				1				
19	2			40	29			2										124		1		1
6	1			14	4													40				
4	2			11	9													36				
5	2			25	7													48		1		1
8				10	5			1										32				
	4		4	9				2	7	11	1							47				
	11	7	7	8				1	13	11	1							89		1		1
	5	6	5	9	2		2	4	3	9								64				
		1		3					1	1								16		1		1
	2	4	1	7	1		1		5	4	1							37		2		2
	3	2	1	3	1	1			4	1								34				
	8	4		5	1		1	3	2	4	1							42				
																		10				
																		20				
																		18				
																		4				
																		14				
																		9				
																		13				
																		26				
																		50				
																		33				
																		22				
																		28				
																		38				
																		38				
															3	3		6			4	4
															7			7			5	5
															5	2		7			2	2
															1			1			2	2
															5	3		8			3	3
																					2	2
																					4	4
													1					1				
																		1				
2	23	16	7	7	2			2		2								95				
4	7	11	14	8	10													74				
1	1	6	1	2	2			1										22				
5	7	13	17	25	4			3										90				
									23	45	2	1						71		5		5
									3	7	3							13		3		3
									13	71	14	2	2					102		12		12
							17											17				
							2											2				
							17											17				
															12	2		14			2	2
															7	1		8			1	1
															33	6		39			10	10
																	1	1				
							2											2				
							1											1				

Anderson - B Species/Elements	Sample 99xxx	451	452	453	454	455	456	457	458	459	460	461	462	463	464	465	466	467	468
Dapsilodus? viruensis	acont.															16	1	3	
	acod.															10	2	1	
Drepanodus arcuatus	arcuat.		1				3		4	2	3	14			1			4	1
	gracili.					1		1				2							
	oist.											2				1		2	1
	pipa.							1	2		1	6						6	1
	sculp.					1	2	1	5		1	9				2		4	2
Drepanodus planus	arcuat.										3				1				1
	gracili.										1								
	oist.										2								
	pipa.					1					1				1				
	sculp.					2									3				
Drepanoistodus basiovalis	suberect.									1	3								
	drep. 2									1	1								
	drep. 3									3	10								
	oist.									1	3								
Drepanoistodus cf. *basiovalis*	suberect.								3							2			
	drep. 1								2										
	drep. 2								3										
	drep. 3								3							4			
	oist.								8							4			
Drepanoistodus forceps	suberect.				1	8	7		1										
	drep. 1					6	1												
	drep. 2		1			5	4		1										
	drep. 3		2		4	37	18	3	4										
	oist.		1		8	33	53	4	6										
Drepanoistodus stougei	suberect.											5							
	drep. 1											7							
	drep. 2											8	2						
	drep. 3											39	2						
	oist.											15	2						
Drepanoistodus tablepointensis	suberect.																		
	drep. 3																		
	oist.																		
"*Drepanoistodus venustus*"	suberect.														1		1		1
	drep. 1																		
	drep. 2																	2	1
	drep. 3																	5	8
	oist.														5	3	3	7	9
"*Drepanoistodus* cf. *venustus*"	suberect.																		
	drep. 1																		
	drep. 2																		
	drep. 3																		
	oist.																		
Eoplacognathus lindstroemi	Pa																		
	Pb																		
Eoplacognathus reclinatus	Pa																		
	Pb																		
Eoplacognathus suecicus	Pa																		
	Pb																		
Eoplacognathus sp. indet.	Pa																		
Erraticodon sp.	P																		
	S																		
Fahraeusodus aff. *marathonensis*	P			1															
Goverdina? sp.	Sc																		

469	470	471	472	473	474	475	476	477	478	479	480	481	484	485	486	488	491	Total	Andersön - C	492	496	Total
7	4	19		6			13	14	8	31	10				56			188			7	7
1	2	7		2		1	4	4	2	9	3	1			10			59			6	6
2	10	1	6	1		1	1	1	4	6						2		68			3	3
									1	1								6				
	2	2														1		11				
	3		2	1			2			2						1		28			1	1
3	14	6	13	7	2	1	2		6	9								90			2	2
7		1	5	1					4	2	2							27				
																		1				
																		2				
											1							4				
	4	2	3	7					1	4								26				
																		4				
																		2				
																		13				
																		4				
																		5				
																		2				
																		3				
																		7				
																		12				
																		17				
																		7				
																		11				
																		68				
																		105				
																		5				
																		7				
																		10				
																		41				
																		17				
				1														1				
				3														3				
				1														1				
	1	3	2				1		4	2								16				
	2	2						1		2						2		9				
		3	4						1	1					1			13		1	2	3
1	16	23	11	1			3	2	5	32			1		4		1	113		1	4	5
3	20	24	5				7	5	5	12			2	2	20			132		1	2	3
															1			1				
															1			1				
															2			2				
															10			10				
															5			5				
															4			4			6	6
															7			7			5	5
													4					4				
													8					8				
									6	9	1							16				
									9	12	4							25				
																				1		1
															2			2				
															1			1				
																		1				
1				5														6				

Andersön - B

Species/Elements	Sample 99xxx	451	452	453	454	455	456	457	458	459	460	461	462	463	464	465	466	467	468
Histiodella holodentata	P																		
Histiodella kristinae	P																		
Lenodus cf. *variabilis*	Pa															1			1
	Pb														3	2		3	
	M																		1
	Sa															1		1	
	Sb														1			1	
	Sd														3	2			1
Lenodus sp. A	Sb						2		2		1								
Lenodus sp. B	Pb												1						
	Sa											3	2						
	Sc											2							
	Sd											2	2						
Lenodus sp. indet.	Pa																		
Microzarkodina flabellum	P					5	57		1										
	M					3	17		1										
	Sa					1	15												
	Sb						7		1										
	Sc					1	24		2										
	Sd						9												
Microzarkodina hagetiana	P														46	17	9		1
	M														50	14	13		1
	Sa														8	1	1		
	Sb														2	2	1		
	Sc														10	5	4		2
	Sd														5		2		
Microzarkodina ozarkodella	P																	1	
	M																		
	Sa																		
	Sb																		
	Sc																		
	Sd																		
Microzarkodina parva	P								1	1	18		5		3				
	M									1	5		4						
	Sa										2								
	Sb										2								
	Sc										8		1						
	Sd										6								
Minimodus poulseni	drep. 1																		
	drep. 2																		
	drep. 3																		
	oist.																		
	scand.																		
Nordiora torpensis	Pa															6			
	Pb																		1
	Sc															1			
Nordiora sp. A	Pa																		
	Pb																		
	M																		
	Sb																		
	Sc																		
	Sd																		
Oelandodus elongatus	S		1	2															
Oistodus lanceolatus	P					2													
Oistodus tablepointensis	S																		
Paltodus subaequalis	drep.	2																	
	oist.		3																

469	470	471	472	473	474	475	476	477	478	479	480	481	484	485	486	488	491	Total	Andersön - C	492	496	Total
																				2		2
									1	128								129				
	1		4				1			5	3							16		1		1
			3				5			13	2							31				
			5															6				
																		2				
			1							2								5				
	2		1															9		1		1
																		5				
																		1				
																		5				
																		2				
																		4				
					1													1				
																		63				
																		21				
																		16				
																		8				
																		27				
																		9				
2		5																80				
																		78				
																		10				
																		5				
1																		22				
																		7				
	6	55	3				1		2									68		8		8
	4	30	2				1		1									38		7		7
		7																7		1		1
	1	2					1											4				
	1	25		2			1		1									30		4		4
	1	22				1												24		1		1
																		28				
																		10				
																		2				
																		2				
																		9				
																		6				
					2				25	3								30				
				4	3		1	2	7	2								19				
				3	3	2	9	4	55	18								94				
				2		1	8	3	36	13	1							64				
									1	3								4				
																		6				
																		1				
																		1				
		1																1				
		2																2				
		1																1				
		3																3				
		1																1				
		2																2				
																		3				
																		2				
					1													1				
																		2				
																		3				

Andersön - B Species/Elements	Sample 99xxx	451	452	453	454	455	456	457	458	459	460	461	462	463	464	465	466	467	468
Panderodus sulcatus	falcif.																		
	arcuatif.																		
Paracordylodus gracilis	P	1		1															
	M	4	4	1															
	S	30	20																
Parapaltodus simplicissimus	drep.															1			
Paroistodus horridus	drep.																1	1	
	oist.																		
Paroistodus originalis	drep.				3	2	3	20	45	5	62	47							1
	oist.					5	5	7	38	6	45	36				1			
Paroistodus parallelus	drep.			2															
	oist.			1															
Paroistodus proteus	drep.	17	10																
	oist.	6	4																
Periodon aculeatus	Pa																		
	Pb																		
	M																		
	Sb																		
	Sc																		
	Sd																		
Periodon flabellum	Sc						2												
Periodon macrodentata	Pa														7	8	2	15	6
	Pb														6	5	3	2	
	M														5	25		19	13
	Sa															3		1	1
	Sb														5	5		2	2
	Sc														4	11	2	3	3
	Sd															4		6	4
Periodon selenopsis	Pa					1													
	M				1	6													
	Sa					1													
	Sb				1	3													
	Sc					5													
Periodon zgierzensis	Pa							3	8	1									
	Pb							5	5										
	M							5	5										
	Sb								6	1									
	Sc							2	8										
	Sd							2											
Polonodus clivosus	Pa															1			
	Pb?															1			
	M															1			
Polonodus cf. *clivosus*	Pa-1																		
	Pa-2																		
	M																		
	Sb																		
Polonodus? cf. *tablepointensis*	Pa-1																4	2	15
	Pa-2																1	3	3
	Pb																		
	M																		2
	Sa																	1	1
	Sb																	1	3
	Sd																		3
Polonodus sp. indet.	Pa-2																		
	Pb																		
	M																		
	Sa																		
	Sb																		
	Sd																		

469	470	471	472	473	474	475	476	477	478	479	480	481	484	485	486	488	491	Total	Andersön - C	492	496	Total
									1									1				
									2									2				
																		2				
																		9				
																		50				
																		1				
	1		1	2														6				
				1														1				
			1	5														194				
				3														146				
																		2				
																		1				
																		27				
																		10				
													8		7			15				
															2			2				
													7		14			21				
													3		1			4				
													8		6			14		1		1
													5		2			7		1		1
																		2				
18	15		7	13	7		2		26	13	35							174		7		7
7	12		7	6	4			1	22	10	17							102		2		2
35	14	17		14	8		4	1	44	7	43							249		9		9
4	7		3	4	2			1	6	4	4							40		1		1
17	8		4	5	4				19	3	23							97		1		1
34	8		18	12	6			1	26	9	9							146		9		9
19	7	1	3	4	3	1	1		24	3	17							97		3		3
																		1				
																		7				
																		1				
																		4				
																		5				
																		12				
																		10				
																		10				
																		7				
																		10				
																		2				
																		1				
																		1				
																		1				
				2					1									3				
				1					3									4				
									4									4				
									2									2				
6	1	12	11	8						4								63				
1			11	1			2											22				
2	5	7	3	13														30				
3				7														12				
	1	1	1							1								6				
	2	5	4	3						1								19				
			1	2						1								7				
								1										1				
			1					1										2				
			2															2				
			1															1				
			1															1				
									1									1				

Andersön - B

Species/Elements	Sample 99xxx	451	452	453	454	455	456	457	458	459	460	461	462	463	464	465	466	467	468
Protopanderodus calceatus	scand.														1	2			1
	acont.				2						3		1		14	14	5	22	12
Protopanderodus graeai	scand.																		
	sym. acont.																		
	asym. acont.																		
Protopanderodus rectus	scand.				1	11	6	2	6		6	30	1						
	sym. acont.				2	43	22	12	20	2	27	75	4	1					
	asym. acont.				2	22	5	9	11		10	56	2						
Protopanderodus robustus	scand.																		14
	sym. acont.																		18
	asym. acont.																		6
Pygodus anserinus	Pa																		
	Pb																		
	Sa																		
	Sb																		
	Sc																		
	Sd																		
Pygodus serra	Pa																		
	Pb																		
	Sb																		
	Sd																		
Scalpellodus gracilis	scand.														6	9		3	2
	longb. drep.														22	54	5	10	7
	shortb. drep.														9	23	3	5	5
Scalpellodus latus	scand.										5	14	3		2				
	longb. drep.							2	1		15	59	18						
	shortb. drep.							2	2	2	6	12	5						
Scalpellodus sp.	longb. drep.													1					
Scolopodus quadratus			2					12											
"*Scolopodus*" *peselephantis*												5	3					2	2
Semiacont. cornuformis	cornuf. A															1			
	cornuf. C														4	5		3	
	drep.													3	10	36	3	14	
Stolodus stola	P			1															
	Sc			1															
	Sd			1			1												
Strachanognathus parvus																3			4
Tetraprioniodus robustus	Pa	7	3																
	Pb	2	2																
	M	2	1																
	Sa	2																	
	Sd	4	2																
Triangulodus brevibasis	P						1	1	2	1									
	M							2	1										
	Sa								1										
	Sb								2	1									
Gen. et sp. indet. A	Pa?		1																
Gen. et sp. indet. B	scand.																		
Gen. et sp. indet. D	Pa															2			
	Pb															3			
unidentifiable		1		1				1			4	1		2	6	4		1	
Total		80	56	12	27	205	269	106	269	45	328	629	92	7	278	362	75	251	185

469	470	471	472	473	474	475	476	477	478	479	480	481	484	485	486	488	491	Total	Andersön - C	492	496	Total
		1	3								1					1		10				
2		1	15						5	7	2				1	1		107			3	3
				4	2	1	2		9	4	3							25				
				31	9	3	19	18	93	56	25					11		265				
				21	3	1	7	18	19	8	15					9		101				
																		63				
																		208				
																		117				
15	16	8	4	32	10	1	16		44	47	11	1			6			225		2		2
26	43	14	29	45	64	2	66		131	123	46	1		2	13			623		7	15	22
29	17	11	9	31	13	2	36		96	137	8	1			6			402		4	8	12
														3	17	68		88			80	80
														1	12	55		68			29	29
															1	8		9				
																18		18			3	3
															1			1				
															2	8		10				
													8					8				
													13					13				
													1					1				
													2					2				
1	2	4	1	3	1			1	3	1								37				
1	9	11	6	7	4			3	6	22								167				
		7	9	7	2	2			1	6								79				
																		24				
																		95				
																		29				
																		1				
																		14				
	2		2	4				2	1	9								32				
																		1				
2		5	1															20				
8		13	3	4														94				
																		1				
																		1				
																		2				
3																		10				
																		10				
																		4				
																		3				
																		2				
																		6				
																		5				
																		3				
																		1				
																		3				
																		1				
							1		2	6								9				
																		2				
																		3				
2	1			6			1			1	2	1		6	2	3		46				
317	350	422	299	529	231	28	258	103	848	967	315	8	74	14	265	230	2	8536		100	213	313

Appendix 4

Distribution of conodont elements in the Herram and Steinsodden sections.

Herram – Steinsodden Species\Element	Sample 69xxx	97792	97796	682	685	687	603	606	608	611	613	614	616	617	619	621	62:
Ansella jemtlandica	P																
	M																
	Sa																
	Sb																
	Sc																
Baltoniodus medius	Pa																
	Pb																
	M																
	Sa																
	Sb																
	Sc																
	Sd																
Baltoniodus navis	Pa			1	6	1	2	1	1				1				
	Pb				8	1	2					1	6				
	M			1	4		4		2			1	3				
	Sa			1	4	1	1		1			1	4				
	Sb				10	1	1		1			1	3				
	Sc				2								3				
	Sd				12	3	3		1		2	1	4				
Baltoniodus norrlandicus	Pa														2		5
	Pb																
	M																3
	Sa																
	Sb													1		1	4
	Sc																
	Sd													1		1	
Coelocerodontus? aff. latus				1													
Cornuodus longibasis		1	1	2	8	5					1		3	1		1	2
Costiconus costatus	drep.					1	1		1		1	1		1		1	
	oist.													3			
	costate				7	4	1	1	1				4	1			3
Costiconus ethingtoni	drep.																
	oist.																
	costate																
Costiconus cf. *ethingtoni*	oist.																
Costiconus sp. *indet.*	costate																
Dapsilodus? viruensis	acont.																
	acod.																
Drepanodus arcuatus	arcuat.	1			3	2	1	2	2		3	2					
	gracili.				1				1								
	oist.			1	1	1											
	pipa.											2			1		
	sculp.			1	4	12	2	3	1					3	1	1	
Drepanodus planus	arcuat.				3												
	pipa.													1			
Drepanoistodus basiovalis	drep. 3											5	1				
	oist.											1	1				
Drepanoistodus aff. *basiovalis*	drep. 3																
	oist.																
Drepanoistodus cf. *basiovalis*	suberect.				3												
	drep. 1				2				2								
	drep. 2				4	2	1	5									
	drep. 3				9	4	5	4									
	oist.				10	4		1									

624	626	628	629	631	633	635	638	642	644	647	649	653	654	656	659	662	667	668	671	673	675	676	677	Total
																13		1						14
																6						1		7
																3								3
											1					7						1		9
																3					2		1	6
							1	2	2	2	3					4		1						15
						5	3	1			6	3				9		3				2		32
						2		2	3	2	8	5			1	8	1	2		1	1	1		37
						1	1			5	5				1	4					1	1		19
								3		3	9				1	5		1		1				23
						3	1	3	1	4	6	3				2	1	1			1			26
						3	4		3	3	7	3	1		2	8							1	35
																								13
																								18
																								15
																								13
																								17
																								5
																								26
3	1	1	3	5	1																			21
4	1		3	11	2																			27
8	2		3	11	2																			29
2	1		1	4	3																			11
2			6		1																			13
1	1		3	13																				21
3	1		4	12																				26
																								1
		2	19	21	4	8	46	7	1	5	8			2		6								159
			16		4	2	2			5	2					5	3							47
			2			3	2			1							1							13
	1	2	49		16	4	11	3	3	11	2					3	1							128
																					8	2		10
																					6			6
																					25	3		28
																			1	1				2
																		1						1
											3							8			4	2		17
											2			1			1	7		1	6	2	1	21
	1	1	2	6	1		3		1		3	1		1				1			4		1	43
			1								1													4
								1																5
											1							1		1				6
	5	1	1	5		1	1		1		1													45
								1			1										1			6
																								1
																								6
		5																						10
					1																			1
					1	4																		5
																								3
																								4
																								12
																								22
																								15

Herram – Steinsodden

Species\Element	Sample 69xxx	97792	97796	682	685	687	603	606	608	611	613	614	616	617	619	621	62
Drepanoistodus forceps suberect.				1	1	1											
drep. 1				2		2											
drep. 2				1	4												
drep. 3				19	18	11											
oist.				17	7	16											
Drepanoistodus stougei suberect.															1	2	1
drep. 1																	9
drep. 2																	4
drep. 3															1	2	8
oist.														2		1	3
Drepanoistodus cf. *stougei* suberect.													2				
drep. 1													1				
drep. 2													2				
drep. 3													6				
oist.													1				
Drepanoistodus tablepointensis oist.																	
"*Drepanoistodus venustus*" suberect.																	
drep. 1																	
drep. 2																	
drep. 3																	
oist.																	
Drepanoistodus sp. *indet.* drep. 1											1						
drep. 3		2	1								1						
Eoplacognathus suecicus Pa																	
Pb																	
Eoplacognathus sp. Pa																	
Pb																	
Fahraeusodus marathonensis P																	
S																	
Histiodella holodentata P																	
Histiodella kristinae P																	
Lenodus cf. *variabilis* Pa																	
Pb																	
M																	
Sa																	
Sb																	
Sc																	
Sd																	
Microzarkodina corpulenta P						16											
M						2											
Sa						1											
Sc						1											
Sd						1											
Microzarkodina flabellum P				14	11	43			2								
M				3	5	5			3								
Sa				4	2	1			1								
Sb				6	4	1											
Sc				7	2	3			2								
Sd						3											
Microzarkodina hagetiana P																	
M																	
Sa																	
Sc																	
Microzarkodina ozarkodella P																	
M																	
Sb																	
Sc																	
Sd																	

624	626	628	629	631	633	635	638	642	644	647	649	653	654	656	659	662	667	668	671	673	675	676	677	Total
																								3
																								4
																								5
																								48
																								40
		2	13			1																		20
1	3	10	17																					40
		1	4	36		1																		46
1	2	19	54			6																		93
1	2	16	81	3		2																		111
																								2
																								1
																								2
																								6
																								1
						2																		2
				1		1	2				3										1	1		9
				3		3			1	2	4	1									2	2		18
				2		1	5	1			1										1			11
				11		4	13	4	2	1	9				2	1	2	5			7	4		65
				14	4	8	12	1	3	3	7							3			3	5		63
																								1
																								4
																				1	1			2
																				1	2			3
																1								1
																	1							1
						1																		1
				1		2																		3
						8	7	2							1		1	13						32
																		2						2
						1		1		1		1												4
			1	1		4	3			5	2				1			4						21
			1		1		2																	• 4
					1		1											1						3
						1																		1
						1	1																	2
				1		1	1																	3
																								16
																								2
																								1
																								1
																								1
																								70
																								16
																								8
																								11
																								14
																								3
							7	2	1		26													36
							13	1	2	1	9													26
											3													3
							1		1		11													13
							3							4	2		1	9						19
														4	3		1	6						14
														2										2
														2	1			4						7
																		1						1

Herram – Steinsodden Species\Element	Sample 69xxx	97792	97796	682	685	687	603	606	608	611	613	614	616	617	619	621	62.
Microzarkodina parva	P									1			2			1	5
	M										1		2		1		4
	Sa																1
	Sb																
	Sc																
Microzarkodina sp. indet.	M																
Minimodus poulseni	drep. 2																
	drep. 3																
	oist.																
	scand.																
Nordiora torpensis	Pa																
	Pb																
	M																
	Sb																
	Sd																
Oepikodus evae	Pa			5	8	18											
	Pb			2	3	7	1										
	M			2	4	5											
	Sb			3	4	7											
	Sc				3	4											
	Sd			3	9	8											
Oistodus tablepointensis	S																
Paltodus deltifer	drep.	1															
	oist.	1															
Paltodus cf. *subaequalis*	drep.	5	1														
	acod.	2	1														
	oist.	2	1														
Paracordylodus gracilis	M			1	2												
	S			2	2												
Paroistodus horridus	drep.																
Paroistodus numarcuatus	drep.	5	2														
	oist.	5															
Paroistodus originalis	drep.			4	145	218	20	2	13	4	3	18	20	8	1	1	35
	oist.			4	68	108	28	2	13	1	1	6	10	4			39
Paroistodus proteus	drep.	3															
	oist.	4															
Periodon aculeatus	Pa																
	Pb																
	M																
	Sa																
	Sb																
	Sc																
	Sd																
Periodon flabellum	Pa			2	6	2											
	Pb			3	6	3											
	M			8	6	6											
	Sa			3	4	2											
	Sb			4	6	2											
	Sc			11	8	5											
	Sd			1	7	3											
Periodon macrodentata	Pa																
	Pb																
	M																
	Sa																
	Sb																
	Sc																
	Sd																

624	626	628	629	631	633	635	638	642	644	647	649	653	654	656	659	662	667	668	671	673	675	676	677	Total
			4																					13
2	1		7	2																				20
																								1
			1																					1
	1		3	1																				5
						2							2											4
																	1			2				3
																	1							1
																	1		1		1			3
																	1							1
											1													1
								6		1														7
										2														2
							2	3		8														13
								1		3														4
																								31
																								13
																								11
																								14
																								7
																								20
																					3			3
																								1
																								1
																								6
																								3
																								3
																								3
																								4
																						1		1
																								7
																								5
4					1					1						1								499
							1																	285
																								3
																								4
																							18	18
																							9	9
																							25	25
																							7	7
																							6	6
																							22	22
																							19	19
																								10
																								12
																								20
																								9
																								12
																								24
																								11
			2	2	2			5	2	5	6				1	1	5	2			36	1		70
					2			1		5	3						2		1		17	1		32
			3					1		4	3					1	3	4			62	5		86
								6		1	2						2				15			26
								1		2							3	2	1		31	1		41
			3	1	1		1	2		5	9						6	3	5		47	2		85
								1		1	2						1	1			15			21

Herram – Steinsodden

Species\Element	Sample 69xxx	97792	97796	682	685	687	603	606	608	611	613	614	616	617	619	621	622
Periodon zgierzensis	Pa				4	2		3						1	1		5
	Pb				1		1	2				1		1			1
	M				1	2	3	2			1	1	1		1		5
	Sa					1		1			1			2			2
	Sb				1	1	1				1						1
	Sc				1	2		3				2	1	3	1		10
	Sd				2	1		2				1		1		1	1
Polonodus? cf. *tablepointensis*	Pa-1																
	Pa-2																
	Pa, undiff.																
	Pb																
	M																
Polonodus? sp. B	Pb																
Polonodus sp. indet.	Pa, undiff.																
	Pb																
	M																
	Sd																
Protopanderodus calceatus	scand.				1			1			1						1
	acont.				6	7		3				1					5
Protopanderodus graeai	scand.																
	sym. acont.																
	asym. acont.																
Protopanderodus rectus	scand.			5	13	54	12	4	4		1	2	5	3			3
	sym. acont.			14	34	85	9	7	12		2	5	6	3	1		6
	asym. acont.			2	24	38	4	2	5	1	2	2	4	1			3
Protopanderodus robustus	scand.																
	sym. acont.																
	asym. acont.																
Protopanderodus sp. indet.	asym. acont.																
Protoprioniodus sp. A	P			2													
	Sb			3													
Scalpellodus gracilis	scand.																
	longb. drep.																
	shortb. drep.																
Scalpellodus latus	scand.																1
	longb. drep.								3			4	3			1	2
	shortb. drep.										1	1	4				1
Scolopodus quadratus						5											
"*Scolopodus*" *peselephantis*		2			2	1			1		1	3	1				3
Semiacont. cornuformis	cornuf. A																
	cornuf. B																
	cornuf. C																
	drep.																
Stolodus stola	P				1												
	Sb					2											
	Sc				1	3											
	Sd			2	4	6											
Strachanognathus parvus																	
Trapezogn. quadrangulum	Pa				5	1											
	Pb			1	2	3											
	M				3	2											
	Sa			1	1	5											
	Sb			3	1												
	Sc			1	1		1										
	Sd				4	2											

624	626	628	629	631	633	635	638	642	644	647	649	653	654	656	659	662	667	668	671	673	675	676	677	Total
2	1	5																						24
2		1																						10
1	1	1																						20
	1																							8
		2																						7
2		7																						32
		1																						10
						1	1				4										1			7
							2																	2
					2					1														3
						1					5										1			7
										1	1													2
																	1							1
								1								2		1						4
								1	3			1						1						6
																1								1
												1												1
							2	1	5	2	9													23
	3		1				2	18	10	16	27										2	3		104
																3	1	3			7		1	15
																14	6	7	1	2	21	2	5	58
																6		4	1	1	7	1	9	29
2		1	3		2	27	6	23	8	9	3													190
2		5	23	12	14	70	20	70	21	33	6													460
	1	4	18	9	5	34	16	32	10	14	5													236
																2	2	2			8	2		16
																28	3	3	2	3	27	2		68
															1	9	1		1	1	9	3		25
																1			1	1	1			4
																								2
																								3
			1	2					1	2	6	2				6	1	1						22
			4	10			2	8	6		23	5	1	4	2	16	1	7	1					90
			2	2			4	1	2	16	3		2	2	6		3		1				44
			2	8	2	2	2		1		1													19
2	4	3	7	26	15	18	3				3													94
4	2	1	5	19	5	6	11																	60
																								5
							6									1		1			3			25
		1	2	5		2																		10
			1	1																				2
							2		2	6	1			1	1	3								16
	2	2	4	9	6	7	3	4	3	4	15	1			1	4	2	1						68
																								1
																								2
																								4
																								12
							2	2																4
																								6
																								6
																								5
																								7
																								4
																								3
																								6

Herram – Steinsodden Species\Element	Sample 69xxx	97792	97796	682	685	687	603	606	608	611	613	614	616	617	619	621	622
Triangulodus amabilis	P				3	10	2										
	M				2	3	1										
	Sa			1	2	2			1								
	Sb			1	3	5			1								
	Sc				1	3			1								
	Sd				1	3											
Gen. et sp. indet. A	Pa?	1															
	Pb?	1															
Gen. et sp. indet. B	oist.																
Gen. et sp. indet. C	P?												2				
	Sd												1				
Unidentifiable		1		8	10	5	4		3			3					2
TOTAL		37	7	184	550	822	115	47	89	7	26	62	113	41	10	13	202

624	626	628	629	631	633	635	638	642	644	647	649	653	654	656	659	662	667	668	671	673	675	676	677	Total
																								15
																								6
																								6
																								10
																								5
																								4
																								1
																								1
																					1			1
																								2
																								1
2			4	18	4	3	1	3	3	2	2	8		1			1	1	4	1		9		2	105
51	39	101	424	245	101	252	236	229	104	176	306	31	5	24	24	214	49	120	9	18	400	51	128	5662

Appendix 5

Distribution of conodont elements in the Røste and Jøronlia sections.

Species / Elements	Sample 97xxx	Røste							Jøronlia				
		757	759	760	763	766	768	Total	718	719	723	730	Total
Ansella jemtlandica	P				1		2	3		2			2
	M						3	3		1			1
	Sb			1			1	2	1				1
	Sc						3	3					
Baltoniodus medius	Pa				1			1		1		1	2
	Pb			1	1		1	3	1	1		1	3
	M			2	5	1	1	9	2			2	4
	Sa			1	1			2	1			1	2
	Sb			2	1	2	1	6	1			3	4
	Sc			1	2		2	5	1				1
	Sd			1	4	1	3	9		1		3	4
Baltoniodus sp. indet.	M		2					2					
	Sa		1					1					
Coelocerodontus? sp.							2	2					
Cornuodus longibasis			1	13	2	3	1	20		1			1
Costiconus costatus	drep.			4	2			6	1				1
	oist.			3	2			5		1			1
	costate			3	5			8		1			1
Costiconus ethingtoni	drep.					1	14	15					
	oist.					1	3	4					
	costate					1	14	15					
Costiconus cf. *ethingtoni*	drep.											2	2
	oist.				1			1				2	2
	costate											2	2
Dapsilodus? viruensis	acont.				1	3	2	6		1		1	2
	acod.					1	2	3	1			2	3
Drepanodus arcuatus	arcuat.		2	4	4		2	12	5				5
	oist.			1				1					
	pipa.		1	3	2		1	7	4				4
	sculp.			4	6		1	11	8				8
Drepanodus planus	arcuat.		2					2		3			3
	pipa.		1					1		1			1
	sculp.		1				1	2					
Drepanoistodus basiovalis	suberect.					1		1					
	drep. 1					1		1					
	drep. 2					5		5					
	oist.					1		1					
"Drepanoistodus venustus"	suberect.			1				1		1			1
	drep. 1			1	1		3	5		1		1	2
	drep. 2											1	1
	drep. 3			5	1	1	3	10		4		3	7
	oist.			4		1	6	11	1	2		2	5
Eoplacognathus suecicus	Pa						1	1					
	Pb						2	2					
Eoplacognathus? sp.	Pa						2	2					
Histiodella sp. A	P											1	1
Lenodus cf. *variabilis*	Pa				1			1	1				1
	Pb		2					2					
Microzarkodina hagetiana	P			12				12					
	M			4				4					
	Sa			2				2					
	Sc			5				5					
	Sd			2				2					

Species / Elements	Sample 97xxx	Røste							Jøronlia				
		757	759	760	763	766	768	Total	718	719	723	730	Total
Microzarkodina ozarkodella	P				1			1					
	M				1			1					
	Sd				1			1					
Microzarkodina sp. indet.	M									2			2
Minimodus poulseni	drep. 1						2	2					
	drep. 2				2		2	4					
	drep. 3						6	6					
	oist.				1		4	5					
Nordiora torpensis	Pa				1	1		2					
	Pb		5		17			22	1			3	4
	M		8		8			16	3				3
	Sa		3		1	3		7					
	Sb		5		6			11	1			1	2
	Sc		2		4			6	2				2
	Sd		7		7	1		15	3			2	5
Paltodus deltifer	drep.	2						2					
	acod.	2						2					
	oist.	2						2					
Paltodus cf. *subaequalis*	oist.	1						1					
Paroistodus horridus	drep.			1			1	2					
Paroistodus numarcuatus	drep.	2						2					
	oist.	1						1					
Paroistodus originalis	drep.					1		1			1		1
Periodon macrodentata	Pa				1	4	3	8	2			1	3
	Pb				1	1	4	6		1		2	3
	M				2	1	5	8					
	Sa					1		1	1				1
	Sb						3						
	Sc				1	4	7	12					
	Sd				2	2	2	6	2				2
Polonodus sp. A	Pa, undiff.											3	3
Polonodus? cf. *tablepointensis*	Pa, undiff.			1	1			2		1			1
	Pb			1	1			2					
Polonodus sp. indet.	Pa, undiff.					1		1					
	M					1		1					
	Sb								2				2
Protopanderodus calceatus	scand.			1	1			2		1		2	3
	acont.		2	4	2			8	7	5		2	14
Protopanderodus graeai	scand.						1	1					
	sym. acont.						4	4					
	asym. acont.						2	2					
Protopanderodus rectus	scand.		2	2	3			7					
	sym. acont.		3	7	13			23					
	asym. acont.		2	12	8			22					
Protopanderodus robustus	scand.						22	22	6	1	1	5	13
	sym. acont.					6	92	98	13	3	1	13	30
	asym. acont.					7	50	57	15	3	1	12	31
Scalpellodus gracilis	scand.			1	2	1	1	5					
	longb. drep.			3	4	1	2	10					
"*Scolopodus*" *peselephantis*				2				2					
Semiacontiodus cornuformis	cornuf. C			1				1					
	drep.			2				2					
Strachanognathus parvus										2			2
Unidentifiable		1	1		1			3	1		2	1	4
Total		11	53	118	137	60	290	669	87	41	6	75	209

Appendix 6

Distribution of conodont elements at Grøslii, Haugnes, Hestekinn, Skogstad and Glöte.

Species/Elements	Sample 97xxx	Grøslii			Haugnes		Hestekinn		Skogstad				Glöte		Total
		664	667	668	674	695	706	715	774	778	785	788	657	660	
Ansella jemtlandica	P				1		1								2
	Sb				1										1
Baltoniodus medius	Pb				1			2							3
	M				1		1	2							4
	Sa							1							1
	Sb					1									1
	Sc				1	1					1				3
	Sd					1	1				1				3
Baltoniodus norrlandicus	Pa				4										4
	Pb				7										7
	M				2										2
	Sa				3										3
	Sb				4										4
	Sc				2										2
	Sd				6										6
Baltoniodus variabilis	Pa													2	2
	M													3	3
	Sd													3	3
Baltoniodus sp. indet.	M												1		1
Cornuodus longibasis					22			3		1			2		28
Costaconus costatus	drep.				6										6
	oist.				1	1									2
	costate				8	1		1							10
Costaconus cf. *dolabellus*	drep.												1		1
	costate												1		1
Dapsilodus? viruensis	acont.							2						2	4
	acod.										1				1
Drepanodus arcuatus	arcuat.				6	1	1	2			1				11
	oist.				2	1		1							4
	pipa.				1										1
	sculp.				6			1			1				8
Drepanodus planus	arcuat.				2								2		4
Drepan. aff. *basiovalis*	oist.				2										2
Drepanoistodus stougei	oist.				2										2

Taxon	Element	1	2	3	4	5	6
"Drepanoistodus venustus"	suberect.	5	3	2			2
	drep. 1	8	4	2			2
	drep. 2	4	2	1	1		
	drep. 3	22	7	1	4	3	7
	oist.	45	11	6	15	3	9
Lenodus? sp. C	Pa	1		1			
	Pb	2		2			
Lenodus sp. indet.	Pa	1					1
	Pb	2					1
Microzarkodina hagetiana	P	4				1	3
	M	2				1	1
Microzarkodina parva	P	11			1		10
	M	7			1		6
	Sa	2					2
	Sd	1				1	
Nordiora torpensis	Pa	1			1		
	Pb	1			1		
	Sb	1			1		
Paltodus cf. *deltifer*	drep.	7					5
	erect.	1					1
	acod.	1					1
	oist.	4					3
Panderodus sp. A	asimilif.	1	1				
	arcuatif.	1	1				
	tortiform ?						
	falcif.	2	2				
	indet.	2	2				
Panderodus sp. B	asimilif.	5	5				
	arcuatif.	2	2				
	falcif.	2	2				
Parapanderodus quietus	sym.	1		1			
	asym.	1		1			
Paroistodus numarcuatus	drep.	5				1	3
	oist.	3				2	1
Periodon aculeatus	M	1	1				1
	Sc	1	1				1
	Sd	1	1				1

Species/Elements	Sample 97xxx	Grøslii			Haugnes		Hestekinn		Skogstad				Glöte		Total
		664	667	668	674	695	706	715	774	778	785	788	657	660	
Periodon macrodentata	Pa				12			1							13
	Pb				12										12
	M				16		1								17
	Sa				7					1					8
	Sb				14		1								15
	Sc				13		1	3							17
	Sd				5		1								6
Polonodus sp. indet.	Pb					1									1
	Sb					1									1
	Sd					1									1
Protopanderodus calceatus	scand.							1							1
	acont.					3		2							5
Protopanderodus rectus	scand.				28	1	1	2							32
	sym. acont.				92	2	4	1	1						100
	asym. acont.				44		3	4	1						52
Protopanderodus robustus	scand.													1	1
	sym. acont.													5	5
	asym. acont.													7	7
Pygodus anserinus	Pa													2	2
	Pb													6	6
Scalpellodus gracilis	longb. drep.					1					3				4
	shortb. drepan.					1					1				2
Scalpellodus sp. indet.	longb. drep.				1		3								4
	shortb. drepan.				1		2						1		4
"Scolopodus" peselephantis			1		1							1			3
Semiacont. cornuformis	drep.		1										5	3	9
Unidentifiable		1	1				1	1	1		1			1	8
Total		13	5	6	381	24	23	61	3	1	11	1	26	83	638

Appendix 7

Distribution of conodont elements at Høyberget, Engerdal and Sorken.

Species/Element	Sample	Høyberget 99510	Høyberget 99511	Engerdal 97655	Engerdal 97656	Sorken 97789	Sorken 97791	Total
Ansella sp. indet.	Sc		1					1
Baltoniodus variabilis	Pb	3						3
	M	1	3					4
	Sa	1						1
	Sd		2					2
Baltoniodus sp. indet.	Pa			1				1
	Pb					2		2
	M			1		1		2
	Sd			1				1
Costiconus sp. indet.	costate				1			1
"*Drepanoistodus venustus*"	suberect,			2				2
	drep. 3	1			2		1	4
	oist.	1	5	8			1	15
Eoplacognathus sp. indet.	Pa				1			1
	Pb			1	2			3
Periodon aculeatus	Pa	1	12					13
	Pb		1					1
	M	1	10					11
	Sc		8					8
	Sd		1					1
Protopanderodus calceatus	scand.				1			1
	acont.		1		6			7
Protopanderodus robustus	scand.					1		1
	sym. acont.	4	1			1		6
	asym. acont.	2				3		5
Pygodus anserinus	Pa		7	5	10			22
	Pb		4	1	4			9
	Sa		1					1
Scabbardella altipes	acod.	10	12	4				26
	distacod.			1				1
	drep.	2	3	1				6
Strachanognathus parvus		1		1				2
Unidentifiable		9	9		1	2	1	22
Total		37	81	27	28	10	3	186